STUDENT SOLUTIONS MANUAL

Bob Martin

PREALGEBRA

SECOND EDITION

Jamie Blair John Tobey
Jeffrey Slater

Prentice Hall

Upper Saddle River, NJ 07458

Senior Acquisitions Editor: Paul Murphy
Editorial Assistant: Eileen Nee
Assistant Managing Editor: John Matthews
Production Editor: Wendy A. Perez
Supplement Cover Manager: Paul Gourhan
Supplement Cover Designer: Joanne Alexandris
Manufacturing Buyer: Ilene Kahn

Prentice Hall

Pearson Education, Inc.
Upper Saddle River, NJ 07458

Printed in the United States of America

10 9 8 7 6 5 4 3 2 1

ISBN 0-13-041611-8

Pearson Education Ltd., *London*
Pearson Education Australia Pty. Ltd., *Sydney*
Pearson Education Singapore, Pte. Ltd.
Pearson Education North Asia Ltd., *Hong Kong*
Pearson Education Canada, Inc., *Toronto*
Pearson Educacíon de Mexico, S.A. de C.V.
Pearson Education—Japan, *Tokyo*
Pearson Education Malaysia, Pte. Ltd.
Pearson Education, *Upper Saddle River, New Jersey*

Contents

Chapter 1

Pretest Chapter 1

1. $7032 = 7000 + 30 + 2$

2. (a) $10 > 6$ (b) $0 < 2$

3. (a) $439 = 440$ to nearest ten
(b) $625 = 600$ to nearest hundred

4. $1203 =$ One thousand, two hundred three

5. A number increased by 5: $x + 5$

6. (a) $2 + (6 + y) = (2 + 6) + y = 8 + y = y + 8$
(b) $7 + (x + 4) = (x + 4) + 7 = x + (4 + 7)$
$= x + 11$

7. $301 + 5923 = 6224$

8. $x + 7\Big|_{x=5} = 5 + 7 = 12$

9. $P = 2l + 2w = 2(3) + 2(2) = 10$ ft

10. The difference of eight and two: $8 - 2$

11. $x - 6\Big|_{x=11} = 11 - 6 = 5$

12. $1820 - 356 = 1464$
check: $356 + 1464 = 1820$

13. $8 \cdot 4 \cdot 0 \cdot y = 0$

14. $6(3x) = 18x$

15. $2(x + 3) = 2x + 2(3) = 2x + 6$

16. The product of eighty and a number:
$80n$

17. $4(3000) = 12{,}000$

18. $7612(31) = 235{,}972$

19. $d = rt = 55(8) = 440$ mi

20. The quotient of fifteen and x: $\frac{15}{x}$

21. $\frac{12}{x}\Big|_{x=0} = \frac{12}{0}$ undefined

22. $279 \div 13 = 21$ R6

$$\begin{array}{r} 21 \\ 13\overline{)279} \\ \underline{26} \\ 19 \\ \underline{13} \\ 6 \end{array}$$

check: $13(21) + 6 = 273 + 6$
$= 279$

23. $\$112 \div 7 = \$16$ each

24. $3 \cdot 3 \cdot 3 \cdot 3 = 3^4$

25. $2^3 = 2 \cdot 2 \cdot 2 = 8$

26. (a) x to the fifth power: x^5
(b) 4 squared: 4^2

27. $18 \div 2 - 3 \cdot 2 = 9 - 6 = 3$

28. $y + 8y + 2y + 6xy = 11y + 6xy$

29. $3(5x) = 30 \Rightarrow 15x = 30 \Rightarrow x = 2$

check: $3(5 \cdot 2) \stackrel{?}{=} 30,\ 30 = 30$

30. $(x + 2) + 6 = 10 \Rightarrow x + (2 + 6) = 10$
$x + 8 = 10 \Rightarrow x = 2$

check: $(2 + 2) + 6 \stackrel{?}{=} 10,\ 10 = 10$

31. $8(40)+12(9)=\$428$

32. $15{,}100+14{,}800\approx 30{,}000$
$12{,}300+13{,}700\approx 26{,}000$
4000 mi more in first two years

1.1 Exercises

1. (a) 8002: Eight thousand two
(b) 802: Eight hundred two
(c) 82: Eighty-two
(d) In the number "eight hundred twenty" the place value of the digit "0" is one

3. (a) In the number 9865 the digit 8 is in the hundreds place.
(b) In the number 9865 the digit 5 is in the ones place.

5. (a) In the number 754,310 the digit 4 is in the thousands place.
(b) In the number 754,310 the digit 7 is in the hundred thousands place.

7. (a) In the number 1,284,073 the digit 1 is in the millions place.
(b) In the number 1,284,073 the digit 0 is in the hundreds place.

9. $4967=4000+900+60+7$

11. $2493=2000+400+90+3$

13. $867{,}301=800{,}000+60{,}000+7000$
$+300+1$

15. 5 hundred-dollar bills, 6 ten-dollar bills, and 2 one-dollar bills.

17. (a) 4 ten-dollar bills and 6 one-dollar bills
(b) 4 ten-dollar bills, 1 five-dollar bill, and 1 one-dollar bill. Answers may vary.

19. \$672.00
Six hundred seventy-two and 00/100

21. $1<7$

23. $4<8$

25. $11>10$

27. $9>0$

29. $2131>1909$

31. $52{,}647<616{,}000$

33. $5>2$

35. $2<5$

37. Expedition XLT < Supercab XLT

39. $45=50$ to nearest ten

41. $661=660$ to nearest ten

43. $16{,}462=16{,}500$ to nearest hundred

45. $823{,}042=823{,}000$ to nearest hundred

47. $38{,}431=38{,}000$ to nearest thousand

49. $12{,}577=13{,}000$ to nearest thousand

51. $5{,}254{,}423=5{,}300{,}000$ to nearest hundred thousand

53. $9{,}007{,}601=9{,}000{,}000$ to nearest hundred thousand

55. $865,000 = 870,000$ mi to nearest ten thousand

57. (a) $25,701 = 26,000$ to nearest thousand
(b) $25,701 = 25,700$ to nearest hundred

59. $16,962 = 17,000$ to nearest hundred

1.2 Exercises

1. $10 + x$: ten plus a number

3. To evaluate $x + 6$ if x is equal to 9, replace x with 6 and then add 9 and 6.

5. A number plus 2: $m + 2$

7. The sum of 5 and y: $5 + y$

9. A number added to three: $m + 3$

11. A number increased by 7: $m + 7$

13. $2 + x = x + 2$

15. $8 + n = n + 8$

17. $216 + 3542 = 3758$

19. $n + 5 = 12$

21. $n + 7 + 3 = n + 10$

23. $6 + 2 + x = 8 + x = x + 8$

25. $x + 0 + 2 = x + 2$

27. $(x + 2) + 1 = x + (2 + 1) = x + 3$

29. $5 + (3 + n) = (5 + 3) + n = 8 + n = n + 8$

31. $(n + 3) + 8 = n + (3 + 8) = n + 11$

33. $(x + 2) + 8 = x + (2 + 8) = x + 10$

35. $(6 + n) + 3 = (n + 6) + 3 = n + (6 + 3)$
$= n + 9$

37. $8 + (1 + x) = (8 + 1) + x = 9 + x = x + 9$

39. $(3 + n) + 6 = (n + 3) + 6 = n + (3 + 6)$
$= n + 9$

41. $(7 + a + 1) + 3 = (a + 7 + 1) + 3$
$= (a + 8) + 3$
$= a + (8 + 3)$
$= a + 11$

43. $(1 + x + 7) + 2 = (x + 1 + 7) + 2$
$= (x + 8) + 2$
$= x + (8 + 2)$
$= x + 10$

45. (a) $y + 7\big|_{y=3} = 3 + 7 = 10$
(b) $y + 7\big|_{y=8} = 8 + 7 = 15$

47. $x + y\big|_{x=2,y=10} = 2 + 10 = 12$

49. $a + b + c\big|_{a=9,b=15,c=12} = 9 + 15 + 12 = 36$

51. $n + m + 13\big|_{n=26,m=44} = 26 + 44 + 13 = 83$

53. (a) $B = x + y + 250$
$= 180 + 12 + 250 = \$442$
(b) $B = x + y + 250$
$= 175 + 10 + 250 = \$435$

55.
$$\begin{array}{r} 36 \\ +\,23 \\ \hline 59 \end{array}$$

57. $\begin{array}{r} 142 \\ +34 \\ \hline 176 \end{array}$

59. $\begin{array}{r} 21 \\ 14 \\ 8 \\ +7 \\ \hline 50 \end{array}$

61. $\begin{array}{r} 105 \\ 8 \\ 133 \\ +98 \\ \hline 344 \end{array}$

63. $236+467+26=729$

65. $397+29+467=893$

67. $7287+273+522=8082$

69. $922,876+54+1287+5000=929,217$

71. $3121+8050+16+163,942=175,129$

73. (a) $152+241=\$393$
(b) $63+121+44=\$228$

75. $2(875)+500+24+35=\$2309$

77. $P=2(11)+2(6)=34$ in.

79. $P=4(4)=16$ ft

81. $P=2(3)+5=11$ in.

83. $P=12+4+6+11+(12+6)+7=58$ ft

85. $P=40+30+135+(30+40)+175+40$
$P=490$ in.

Cumulative Review Problems

87. In the number 20,891 the digit 8 is in the hundreds place.

89. $408=400+8$

91. $102,876=103,000$ to the nearest thousand

1.3 Exercises

1. $6-x$: Six minus x

3. The key phrase "how many more" indicates the operation <u>subtraction</u>.

5. $6-2=4$

7. $9-5=4$

9. $9-7=2$

11. $8-7=1$

13. $18-0=18$

15. $20-20=0$

17. $600-500=100$
$600-501=99$
$600-502=98$
$600-503=97$
$600-504=96$

19. $300-200=100$
$300-201=99$
$300-202=98$
$300-203=97$
$300-204=96$
$300-205=95$

21. Seven decreased by three: $7-3$

23. The difference of 8 and *y*: $8-y$

25. Ten subtracted from seventeen: $17-10$

27. Fifteen minus a number: $15-x$

29. Two less than some number: $a-2$

31. $9-n\big|_{n=0}=9-0=9$

33. $9-n\big|_{n=8}=9-8=1$

35. $x+3\big|_{x=3}=3+3=6$

37. $x+3\big|_{x=7}=7+3=10$

39. $8-y\big|_{y=3}=8-3=5$

41. $87-46=41$, check: $46+41=87$

43. $69-34=35$, check: $34+35=69$

45. $83-67=16$, check: $67+16=83$

47. $759-613=146$, check: $613+146=759$

49. $966-177=789$, check: $177+789=966$

51. $500-43=457$, check: $43+457=500$

53. $9922-2667=7255$
check: $2667+7255=9922$

55. $5301-185=5116$
check: $185+5116=5301$

57. $\begin{array}{r} 17{,}008 \\ -\ 4{,}839 \\ \hline 12{,}169 \end{array}$ check: $\begin{array}{r} 4{,}839 \\ +\ 12{,}169 \\ \hline 17{,}008 \end{array}$

59. $\begin{array}{r} 164{,}300 \\ -58{,}923 \\ \hline 105{,}377 \end{array}$ check: $\begin{array}{r} 58{,}923 \\ +105{,}377 \\ \hline 164{,}300 \end{array}$

61. $1999-1877=122$ years

63. (a) Blue: $275{,}000-5{,}000=270{,}000$
Bowhead: $60{,}000-8500=51{,}500$
Humpback: $150{,}000-20{,}000=130{,}000$
The Blue Whale had the largest decline.
(b) The total decline is
$270{,}000+51{,}500+130{,}000=451{,}500$

65. $7900-2160=5740$ mi

67. $100-56=44$ mph slower

69. $400-115=285$ ft more

71. $x-y=y-x$ when *x* and *y* are equal

Cumulative Review Problems

73. $2{,}386{,}702>117{,}401$

75. $31{,}007+579=31{,}586$

77. $430+32+12+28+6=\$508$

1.4 Exercises

1. (a) $4x$: Four times a number
(b) *ab*: The product of *a* and *b*

3. Two times three:

```
          *  *
*  *  *
          *  *
*  *  *
          *  *
```

5. (a)

	White	Pink	Blue
Brown	Brown White	Brown Pink	Brown Blue
Black	Black White	Black Pink	Black Blue
Gray	Gray White	Gray Pink	Gray Blue
Dark Blue	Dark Blue White	Dark Blue Pink	Dark Blue Blue

(b) $4(3) = 12$ different outfits

7. $8(5) = 40$ different dishes

9. $5(7) = 35$, 5,7: factors; 35: product

11. $7a = 28$, 7, a: factors; 28: product

13. A number times 7: $x(7) = 7x$

15. The product of nine and a number: $9x$

17. The product of a and b: ab

19. If $a \cdot b = 0$ and $a = 2$, then $b = 0$

21. If $b(a \cdot c) = 30$, then $(a \cdot b) \cdot c = 30$

23. $(4)(5)(2)(2) = 20(4) = 80$

25. $(5)(4)(3)(2) = 20(6) = 120$

27. $9 \cdot 0 \cdot 3 \cdot 8 = 0$

29. $3 \cdot 2 \cdot 4 \cdot 5 = 6 \cdot 20 = 120$

31. $7(3000) = 21{,}000$

33. $4000 \cdot 8 = 32{,}000$

35. $800 \cdot 8b = 6400b$

37. $500(4z) = 2000z$

39. $7(5b) = 35b$

41. $4(x \cdot 6) = 4(6x) = 24x$

43. $3(9 \cdot c) = 27c$

45. $5(8 \cdot x) = 40x$

47. $5(3)(2 \cdot z) = 15(2z) = 30z$

49. $0(5)(z \cdot 9) = 0$

51. $7(4)(x \cdot 1) = 28x$

53. $6 \cdot 3(5y) = 18(5y) = 90y$

55. $(2a)5 \cdot 4 = (2a)20 = 20(2a) = 40a$

57. $4 \cdot 5(3a)(4b) = 20(12ab) = 240ab$

59. $2(x+1) = 2x + 2(1) = 2x + 2$

61. $6(n-4) = 6n - 6(4) = 6n - 24$

63. $4(x-3) = 4x - 4(3) = 4x - 12$

65. $5(x+9) = 5x + 5(9) = 5x + 45$

67. $7(14) = 7(4+10) = 7(4) + 7(10)$
$= 28 + 70 = 98$

69. $4(103) = 4(3+100) = 4(3) + 4(100)$
$= 12 + 400 = 412$

Cumulative Review Exercises

71. $426,862 + 2128 = 428,990$

73. $826,540 = 827,000$ to nearest thousand

75. $120 - 97 = \$23$ less

1.5 Exercises

1. The distributive property allows you to rewrite 16 as the sum $6+10$ and then mentally multiply $6 \cdot 6 + 6 \cdot 10$.

3. $9(637) = 5733$

5. $7(602) = 4214$

7.
$$\begin{array}{r} 398 \\ \times\ \ 300 \\ \hline 119,400 \end{array}$$

9.
$$\begin{array}{r} 793 \\ \times\ \ 600 \\ \hline 475,800 \end{array}$$

11.
$$\begin{array}{r} 76 \\ \times 68 \\ \hline 608 \\ 456 \\ \hline 5168 \end{array}$$

13.
$$\begin{array}{r} 99 \\ \times 94 \\ \hline 396 \\ 891 \\ \hline 9306 \end{array}$$

15. $56(847) = 47,432$

17. $455(86) = 39,130$

19. $762(309) = 235,458$

21. $409(432) = 176,688$

23. $8324(922) = 7,674,728$

25. $2009(651) = 1,307,859$

27. $12,107(808) = 9,782,456$

29. $61,711(1000) = 61,711,000$

31. $8x\big|_{x=7} = 8(7) = 56$

33. $8x\big|_{x=5} = 8(5) = 40$

35. $7(4y)\big|_{y=2} = 7(4 \cdot 2) = 7(8) = 56$

37. $7(4y)\big|_{y=9} = 7(4 \cdot 9) = 7(36) = 252$

39. $2xy\big|_{x=3,y=7} = 2(3)(7) = 6(7) = 42$

41. $2xy\big|_{x=9,y=8} = 2(9)(8) = 18(8) = 144$

43. $8(40) = \$320$

45. $15(25) = 375$ trees

47. $327(12) = \$3924$

49. $15(60) = 900$ rooms
$50(20) = 1000$ curtains, yes

51. (a) 20 °F (b) $4(19) = 76$ °F

Cumulative Review Problems

53. $8(2)(x \cdot 4) = 16(4x) = 64x$

55. $x + 6 = 8 \Rightarrow x = 2$

1.6 Exercises

1. $220 \div 4$

3. $320 \div 16$

5. $32 \div 4$

7. Twenty-seven divided by x: $27 \div x$

9. Forty-two dollars divided equally among six people: $42 \div 6$

11. The quotient of thirty-six and three: $36 \div 3$

13. The quotient of three and thirty-six: $3 \div 36$

15. $x \div y\big|_{x=36,\,y=36} = 36 \div 36 = 1$

17. $x \div y\big|_{x=66,\,y=66} = 66 \div 66 = 1$

19. $x \div y\big|_{x=14,\,y=0} = 14 \div 0$ undefined

21. $\dfrac{m}{n}\bigg|_{m=24,\,n=12} = \dfrac{24}{12} = 2$

23. $\dfrac{m}{n}\bigg|_{m=0,\,n=40} = \dfrac{0}{40} = 0$

25. $50 \div 6 = 8$ R2

$$\begin{array}{r} 8 \\ 6\overline{)50} \\ \underline{48} \\ 2 \end{array} \qquad \text{check: } 6(8) + 2 = 50$$

27. $2597 \div 7 = 371$

$$\begin{array}{r} 371 \\ 7\overline{)2597} \\ \underline{21} \\ 49 \\ \underline{49} \\ 7 \\ \underline{7} \\ 0 \end{array} \qquad \text{check: } 7(371) = 2597$$

29. $2097 \div 4 = 524$ R1

$$\begin{array}{r} 524 \\ 4\overline{)2097} \\ \underline{20} \\ 9 \\ \underline{8} \\ 17 \\ \underline{16} \\ 1 \end{array} \qquad \text{check: } 4(524) + 1 = 2097$$

31. $1268 \div 30 = 42 \text{ R}8$

$$\begin{array}{r} 42 \\ 30\overline{)1268} \\ \underline{120} \\ 68 \\ \underline{60} \\ 8 \end{array} \quad \text{check: } 30(42)+8=1268$$

33. $632 \div 30 = 21 \text{ R}2$

$$\begin{array}{r} 21 \\ 30\overline{)632} \\ \underline{60} \\ 32 \\ \underline{30} \\ 2 \end{array} \quad \text{check: } 30(21)+2=632$$

35. $5817 \div 19 = 306 \text{ R}3$

$$\begin{array}{r} 306 \\ 19\overline{)5817} \\ \underline{57} \\ 117 \\ \underline{114} \\ 3 \end{array} \quad \text{check: } 19(306)+3=5817$$

37. $2093 \div 41 = 51 \text{ R}2$

$$\begin{array}{r} 51 \\ 41\overline{)2093} \\ \underline{205} \\ 43 \\ \underline{41} \\ 2 \end{array} \quad \text{check: } 41(51)+2=2093$$

39. $1369 \div 19 = 72 \text{ R}1$

$$\begin{array}{r} 72 \\ 19\overline{)1369} \\ \underline{133} \\ 39 \\ \underline{38} \\ 1 \end{array} \quad \text{check: } 19(72)+1=1369$$

41. $18{,}985 \div 27 = 703 \text{ R}4$

$$\begin{array}{r} 703 \\ 27\overline{)18985} \\ \underline{189} \\ 85 \\ \underline{81} \\ 4 \end{array} \quad \text{check: } 27(703)+4=18{,}985$$

43. $11{,}571 \div 34 = 340 \text{ R}11$

$$\begin{array}{r} 340 \\ 34\overline{)11571} \\ \underline{102} \\ 137 \\ \underline{136} \\ 11 \end{array} \quad \text{check: } 34(340)+11=11{,}571$$

45. $113{,}317 \div 223 = 508 \text{ R}33$

$$\begin{array}{r} 508 \\ 223\overline{)113317} \\ \underline{1115} \\ 1817 \\ \underline{1784} \\ 33 \end{array}$$

check: $223(508)+33=113{,}317$

47. $70{,}141 \div 36 = 515 \text{ R}101$

$$\begin{array}{r} 515 \\ 136\overline{)70141} \\ \underline{680} \\ 214 \\ \underline{136} \\ 781 \\ \underline{680} \\ 101 \end{array}$$

check: $136(515) + 101 = 70{,}141$

49. $60 \div 14 = 4 \text{ R}4$
4 tickets were donated to the PTA.

51. $85 \div 5 = \$17$ each

53. $1050 \div 6 = \$175$ per day

55. $124 \div 2 = 62$ ft

57. (a) $36 \div 9 = 4$ babies
(b) $3600 \div 9 = 400$ babies
(c) $3600 \div 17 = 211 \text{ R}13$
212 people

Cumulative Review Exercises

59. Seven plus x equals eleven: $7 + x = 11$

61. $4031(202) = 814{,}262$

63. $1389 - 430 - 495 = 464$ miles

1.7 Exercises

1. What number squared is equal to 16?

3. $3 \cdot 3 \cdot 3 \cdot 3 = 3^4$

5. $a \cdot a \cdot a \cdot a \cdot a = a^5$

7. $8 = 8^1$

9. $2 \cdot 2 = 2^2$

11. $5 \cdot 5 \cdot a \cdot a \cdot a = 5^2 a^3$

13. $3 \cdot 3 \cdot z \cdot z \cdot z \cdot z \cdot z = 3^2 z^5$

15. $7 \cdot 7 \cdot 7 \cdot y \cdot y \cdot x \cdot x = 7^3 y^2 x^2$

17. $n \cdot n \cdot n \cdot n \cdot n \cdot 9 \cdot 9 = n^5 9^2$

19. (a) $y^3 = y \cdot y \cdot y$
(b) $7^5 = 7 \cdot 7 \cdot 7 \cdot 7 \cdot 7$

21. $2^3 = 2 \cdot 2 \cdot 2 = 8$

23. $5^2 = 5 \cdot 5 = 25$

25. $1^6 = 1 \cdot 1 \cdot 1 \cdot 1 \cdot 1 \cdot 1 = 1$

27. $7^2 = 7 \cdot 7 = 49$

29. $4^4 = 4 \cdot 4 \cdot 4 \cdot 4 = 256$

31. $10^1 = 10$

33. $5^3 = 5 \cdot 5 \cdot 5 = 125$

35. $10^5 = 10 \cdot 10 \cdot 10 \cdot 10 \cdot 10 = 100{,}000$

37. $x^2 = (5)^2 = 25$

39. $a^4\big|_{a=1} = 1^4 = 1$

41. Seven to the third power: 7^3

43. Nine squared: 9^2

45. $2 \cdot 4 - 1 = 8 - 1 = 7$

47. $7^2 + 5 - 3 = 49 + 2 = 51$

49. $5 \cdot 3^2 = 5 \cdot 9 = 45$

51. $2 \cdot 2^2 = 2 \cdot 4 = 8$

53. $5^2 - 7 + 3 = 25 - 4 = 21$

55. $9 + 2 \cdot 2 = 9 + 4 = 13$

57. $6^2 \div 6 \times 2 + 1 = 36 \div 6 \times 2 + 1$
$= 6 \times 2 + 1 = 12 + 1 = 13$

59. $8 + (7 + 4^3) = 8 + (7 + 64) = 8 + 71 = 79$

61. $2 \times 15 \div 5 + 10 = 30 \div 5 + 10 = 6 + 10 = 16$

63. $7 + 5(3 \cdot 4 + 7) - 2 = 7 + 5(12 + 7) - 2$
$= 7 + 5(19) - 2$
$= 7 + 95 - 2$
$= 102 - 2 = 100$

65. $59 - 4(1 + 5 \cdot 2) + 4 = 59 - 4(1 + 10) + 4$
$= 59 - 4(11) + 4$
$= 59 - 44 + 4$
$= 15 + 4 = 19$

67. $32 \cdot 6 - 4(4^3 - 5 \cdot 2^2) + 3$
$= 192 - 4(64 - 5 \cdot 4) + 3$
$= 192 - 4(64 - 20) + 3$
$= 192 - 4(44) + 3$
$= 192 - 176 + 3 = 16 + 3 = 19$

69. (a) $(32 \div 4) \div 2 = 8 \div 2 = 4$
(b) $32 \div (4 \div 2) = 32 \div 2 = 16$

71. (a) $(16 \div 4) \div 2 = 4 \div 2 = 2$
(b) $16 \div (4 \div 2) = 16 \div 2 = 8$

73. $2^2 + 8 \div 4 = 4 + 2 = 6$

75. $\frac{(4 + 4 \div 2)}{(5 - 3)} = \frac{4 + 2}{2} = \frac{6}{2} = 3$

77. $\frac{(12 - 2)}{(25 \div 5 \times 2)} = \frac{10}{5 \times 2} = \frac{10}{10} = 1$

79. $2^8 = 256$

81. He should have multiplied 3 times 2 first and then added 4 to get 10.

83. $21 \cdot 10^1 = 210,\ 21 \cdot 10^2 = 2100$
$21 \cdot 10^3 = 21{,}000,\ 21 \cdot 10^4 = 210{,}000$
The exponent on 10 determines the number of trailing zeros attached to the number.

Cumulative Review Problems

85. $4079 + 2762 = 6841$

87. $(387)(196) = 75{,}852$

89. $15(140) = 2100 < 2200$, no.

1.8 Exercises

1. $7x$: seven times x or the product of seven and x

3. $8x = 40$: eight times what number equals 40

5. Two x's: $2x$

7. $a + a + a = 3a$

9. In $5x+3y+2x+8m+6y$, $5x$ and $2x$ are like terms; $3y$ and $6y$ are like terms.

11. In $2mn+3y+4mn+2$, $2mn$ and $4mn$ are like terms.

13. $7x-x=6x$

15. $9y-3y=6y$

17. $3x+2x+6x=11x$

19. $8x+4a+3x+2a=11x+6a$

21. $6xy+4b+3xy=9xy+4b$

23. $6xy+3x+9+9xy=15xy+3x+9$

25. Four plus what number equals sixteen: $4+x=16$

27. If a number is subtracted from forty-five the result is six: $45-x=6$

29. What number times three equals thirty-six: $x\cdot 3=3x=36$

31. Twenty-five divided by what number is equal to five: $\frac{25}{n}=5$

33. $J+10=25$

35. $C-50=1480$

37. $8-x\big|_{x=4}=8-4=4\neq 3$, no

39. $x+4\big|_{x=15}=15+4=19$, yes

41. $x+3=9,\ x=6$

check: $6+3\stackrel{?}{=}9,\ 9=9$

43. $9-n=7,\ n=2$

check: $9-2\stackrel{?}{=}7,\ 7=7$

45. $x-6=0,\ x=6$

check: $6-6\stackrel{?}{=}0,\ 0=0$

47. $1+x=13,\ x=12$

check: $1+12\stackrel{?}{=}13,\ 13=13$

49. $21-x=20,\ x=1$

check: $21-1\stackrel{?}{=}20,\ 20=20$

51. $44-n=42,\ n=2$

check: $44-2\stackrel{?}{=}42,\ 42=42$

53. $8x=16,\ x=2$

check: $8(2)\stackrel{?}{=}16,\ 16=16$

55. $4y=12,\ y=3$

check: $4(3)\stackrel{?}{=}12,\ 12=12$

57. $5x=20,\ x=4$

check: $5(4)\stackrel{?}{=}20,\ 20=20$

59. $10y=30,\ y=3$

check: $10(3)\stackrel{?}{=}30,\ 30=30$

61. $\frac{30}{x}=15,\ x=2$

check: $\frac{30}{2}\stackrel{?}{=}15,\ 15=15$

63. $\frac{14}{x}=2,\ x=7$

check: $\frac{14}{7}\stackrel{?}{=}2,\ 2=2$

65. $\frac{12}{x}=1,\ x=12$

check: $\frac{12}{12}\stackrel{?}{=}1,\ 1=1$

67. $\frac{x}{2}=10,\ x=20$

check: $\frac{20}{2}\stackrel{?}{=}10,\ 10=10$

69. $(x+1)+3=7,\ x+(1+3)=7$
$x+4=7,\ x=3$

check: $(3+1)+3\stackrel{?}{=}7,\ 4+3\stackrel{?}{=}7,\ 7=7$

71. $2+(5+a)=14,\ (2+5)+a=14$
$7+a=14,\ a=7$

check: $2+(5+7)\stackrel{?}{=}14,\ 14=14$

73. $(3+x)+2=7,\ (x+3)+2=7$
$x+(3+2)=7,\ x+5=7,\ x=2$

check: $(3+2)+2\stackrel{?}{=}7,\ 7=7$

75. $3+(n+5)=10,\ (n+5)+3=10$
$n+(5+3)=10,\ n+8=10,\ n=2$

check: $3+(2+5)\stackrel{?}{=}10,\ 10=10$

77. $3n+n=12,\ 4n=12,\ n=3$

check: $3(3)+3\stackrel{?}{=}12,\ 12=12$

79. $9x-5x=20,\ 4x=20,\ x=5$

check: $9(5)-5(5)\stackrel{?}{=}20,\ 20=20$

81. $3y+y+2y=12,\ 6y=12,\ y=2$

check: $3(2)+2+2(2)\stackrel{?}{=}12,\ 12=12$

83. Four plus what number equals eight?
(a) $4+x=8$
(b) $x=4$

85. Seventeen minus what number is equal to seven?
(a) $17-x=7$
(b) $x=10$

87. Triple what number times 2 equals twelve?
(a) $3x(2)=12$
(b) $6x=12,\ x=2$

89. (a) 30 miles per hour
(b) $50=2x,\ x=25$
speed of elephant: 25 miles per hour

91. $2(50)+x=170,\ 100+x=170,\ x=70$
x is 70 ft

93. d

95. a

1.9 Exercises

1. (a) $80+40+20+10=\$150$
(b) $81+36+22+14=\$153$
(c) yes

3. First two days: $600+400=1000$
Second two days: $400+400=800$
$1000-800=200$ mi more the first two days.

5. $2(13)+4(5)+7=\$53$

7. $7439 - 7425 = 14$ for Apt. 100
$8702 - 8671 = 31$ for Apt. 101
$6551 - 6539 = 12$ for Apt. 102
$7179 - 7126 = 53$ for Apt. 103
(a) Apt. 103 used the most gas.
(b) Apt. 102 used the least gas.

9. $\frac{7(475) + 20(65) + 2(650)}{5} = \$1185$

11. $20 + 25 + 80 + 150 + 80 + 25 + 20 = 400$ ft

13. $40(8) + 12(12) = \$464$

15. $35(7) = 245$ gal in one day
$35(7)(7) = 1715$ gal in one week

17. (a) $782,535 - 600,333 = \$182,202$
(b) $182,202 \div 2 = \$91,101$

19. $15(12)(250) = 45,000$ lb

21. $2(7)(4)(6) = \$336 < \$400$
Cheaper to pay each time he rides.

23. $\frac{2(450) + 2(5) + 200}{4} = \$300$

25. (a) $6000 \div 2 = \$3000$
(b) $(6000 \div 2) \div 3 = \$1000$

27. (a) $30 + 240 + 170 = 440$
$440 \div 50 = 8 \text{ R}40 \rightarrow 8(5) = 40$ pts
$40 + 25(240 > 200) \rightarrow 65$ pts total
(b) $65 \div 10 = 6 \text{ R}5 \rightarrow \$6$ in discounts

29. (a) $170 + 260 = 430$
$430 \div 50 = 8 \text{ R}30 \rightarrow 8(10) = 80$ pts
$80 + 50(260 > 200) \rightarrow 130$ pts
(b) $130 \div 25 = 5 \text{ R}5,\ 5 \cdot \$5 = \$25$ discount

31. (a) $440,000 + 140,000 + 130,000$
$+210,000 = \$920,000$
(b) $435,750 + 144,800 + 125,500$
$+213,125 = \$919,175$
(c) yes

33. $175(36) - 5000 = \$1300$

Cumulative Review Problems

35. $215 \div 5 = 43$

37. $x - 22 = 36,\ x = 58$

Putting Your Skills to Work, Individual Investigation and Analysis

1. Plan 1:
$4(450) + 13(40) + 13(80)$
$+13(75) = \$4335$
Plan 2:
$350(2) + 175(2) + 2(3)(40)$
$+12(75) + 17(80) = \$3550$
Plan3:
$22(20) + 21(75) + 21(80) = \$3695$

2. Plan 2 is cheapest.

3. $3550 + 4(65) = \$3810$
Plan 3 is now cheapest.

Putting Your Skills to Work, Group Investigation and Cooperative Problem Solving

1. Plan 1 is still \$4335 since no additional vacation days are required. Plan 2 is now $3550 + 240 = \$3790$ since it requires one additional vacation day. Plan 3 is now $3695 + 5(240) = \$4895$ since it requires one additional vacation day.

3. Plan 2 is now cheapest.

Chapter 1 Review

1. (a) In the number 175,493 the digit 7 is in the ten thousands place.
(b) In the number 175,493 the digit 5 is in the thousands place.

2. (a) In the number 458,013 the digit 8 is in the thousands place.
(b) In the number 458,013 the digit 0 is in the hundreds place.

3. $7694 = 7000 + 600 + 90 + 4$

4. $4325 = 4000 + 300 + 20 + 5$

5. Three hundred forty-one and 00/100

6. One hundred eighty-seven and 00/100

7. $2 < 8$

8. $12 > 11$

9. $11 < 12$

10. Six is greater than one: $6 > 1$

11. Three is less than five: $3 < 5$

12. $61,269 = 61,300$ to nearest hundred

13. $382,240 = 382,200$ to nearest hundred

14. $6,365,534 = 6,400,000$ to nearest hundred thousand

15. $8,118,701 = 8,100,000$ to nearest hundred thousand

16. Seven more than x: $7 + x$

17. The sum of some number and five: $n + 5$

18. A number increased by four: $x + 4$

19. $7 + (9 + x) = (7 + 9) + x = 16 + x = x + 16$

20. $(n + 4) + 1 = n + (4 + 1) = n + 5$

21. $(2 + n) + 9 = (n + 2) + 9 = n + (2 + 9)$
$= n + 11$

22. $5 + (n + 2) = (n + 2) + 5 = n + (2 + 5)$
$= n + 7$

23. $(5 + x + 3) + 2 = (x + 5 + 3) + 2$
$= (x + 8) + 2 = x + (8 + 2)$
$= x + 10$

24. $(3 + x + 4) + 1 = (x + 3 + 4) + 1$
$= (x + 7) + 1$
$= x + (7 + 1) = x + 8$

25. $7 + n\big|_{n=6} = 7 + 6 = 13$

26. $x + 3\big|_{x=17} = 17 + 3 = 20$

27. $x + y\big|_{x=11, y=15} = 11 + 15 = 26$

28.
$$\begin{array}{r} 63 \\ +\ 32 \\ \hline 95 \end{array}$$

29.
$$\begin{array}{r} 1{,}672{,}045 \\ +\ \ 89{,}204 \\ \hline 1{,}761{,}249 \end{array}$$

30. $8398 + 372 + 255 = 9025$

31. $17,456+213+982=18,651$

32. $1434+1596+1423+1565=6018$

33. $13+5+7+(5+8)+(13+7)+8=66$ m

34. Eight decreased by 3: $8-3$

35. The difference of three and y: $3-y$

36. Ten subtracted from a number: $x-10$

37. $8-x\big|_{x=3}=8-3=5$

38. $a-b\big|_{a=19,b=7}=19-7=12$

39. $7-x\big|_{x=2}=7-2=5$

40. $y-8\big|_{y=9}=9-8=1$

41. $\begin{array}{r} 85 \\ -\ 23 \\ \hline 62 \end{array}$ check: $62+23=85$

42. $\begin{array}{r} 96 \\ -\ 33 \\ \hline 63 \end{array}$ check: $63+33=96$

43. $\begin{array}{r} 763 \\ -\ 219 \\ \hline 544 \end{array}$ check: $544+219=763$

44. $\begin{array}{r} 841 \\ -\ 316 \\ \hline 525 \end{array}$ check: $525+316=841$

45. $8502-2957=5545$
check: $5545+2957=8502$

46. $9021-5862=3159$
check: $3159+5862=9021$

47. $29,104-4988=24,116$
check: $24,116+4988=29,104$

48. $137,405-6877=130,528$
check: $130,528+6877=137,405$

49. $900,000-522,000=\$378,000$ more

50. $720,000-450,000=\$270,000$ less

51. $4x=32$, factors: 4, x; product: 32

52. $xy=z$, factors: x, y; product: z

53. Six times a number: $6x$

54. $7y=63$, Seven times what number equals sixty-three?

55. $7\cdot 2\cdot 3\cdot 0=0$

56. $5\cdot 3\cdot 2\cdot 2=15\cdot 4=60$

57. $2(y\cdot 7)=2(7y)=14y$

58. $3(7)(x\cdot 2)=21(2x)=42x$

59. $2(5)(y\cdot 7)=10(7y)=70y$

60. $(800)6n=4800n$

61. $2(x+1)=2x+2(1)=2x+2$

62. $4(x+1)=4x+4(1)=4x+4$

63. $572(71)=40,612$

64. $\begin{array}{r} 406 \\ \times \quad 32 \\ \hline 12{,}992 \end{array}$

65. $(4251)352 = 1{,}496{,}352$

66. $\begin{array}{r} 6424 \\ \times \quad 903 \\ \hline 5{,}800{,}872 \end{array}$

67. $\begin{aligned} 5(8y)\big|_{y=10} &= 5(8 \cdot 10) \\ &= 5(80) \\ &= 400 \end{aligned}$

68. $\begin{aligned} 6mn\big|_{m=8,n=2} &= 6 \cdot 8 \cdot 2 \\ &= 96 \end{aligned}$

69. $17(18) = 306$ mi

70. $6(21)(4) = 504$ doors

71. $300 \div 20$

72. $\$500 \div n$

73. $5 \div y$

74. $26 \div 13$

75. $\left.\frac{x}{y}\right|_{x=10,y=0} = \frac{10}{0}$ undefined

76. $\begin{aligned} \left.\frac{a}{b}\right|_{a=33,b=33} &= \frac{33}{33} \\ &= 1 \end{aligned}$

77. $\begin{array}{r} 451 \\ 4\overline{)1804} \\ \underline{16} \\ 20 \\ \underline{20} \\ 4 \\ \underline{4} \\ 0 \end{array}$

78. $\begin{array}{r} 243 \\ 7\overline{)1701} \\ \underline{14} \\ 30 \\ \underline{28} \\ 21 \\ \underline{21} \\ 0 \end{array}$

79. $2485 \div 31 = 80$ R5

$\begin{array}{r} 80 \\ 31\overline{)2485} \\ \underline{248} \\ 5 \end{array}$

80. $1456 \div 29 = 50$ R6

$\begin{array}{r} 50 \\ 29\overline{)1456} \\ \underline{145} \\ 6 \end{array}$

81. $369{,}757 \div 922 = 401$ R35

$\begin{array}{r} 401 \\ 922\overline{)369757} \\ \underline{3688} \\ 957 \\ \underline{922} \\ 35 \end{array}$

82. $\frac{510{,}144}{846} = 603 \text{ R6}$

$$\begin{array}{r} 603 \\ 846\overline{)510144} \\ \underline{5076} \\ 2544 \\ \underline{2538} \\ 6 \end{array}$$

83. $447 \div 4 = 111 \text{ R3} \to \$3$ deposit

84. $3528 \div 24 = \$147$

85. $2 \cdot 2 \cdot 2 \cdot n \cdot n = 2^3 n^2$

86. $z \cdot z \cdot z \cdot z \cdot 5 \cdot 5 \cdot 5 = z^4 \cdot 5^3$

87. $a^4 = a \cdot a \cdot a \cdot a$

88. $6^5 = 6 \cdot 6 \cdot 6 \cdot 6 \cdot 6$

89. $10^3 = 10 \cdot 10 \cdot 10 = 1000$

90. $2^5 = 2 \cdot 2 \cdot 2 \cdot 2 \cdot 2 = 32$

91. $9^2 = 9 \cdot 9 = 81$

92. Four squared: 4^2

93. x to the fifth power: x^5

94. $6 + 14 \div 2 - 2^2 = 6 + 7 - 4 = 9$

95. $(15 + 25 \div 5) \div (8 - 4) = (15 + 5) \div 4$
$= 20 \div 4 = 5$

96. $2^2 \cdot 5 = 4 \cdot 5 = 20$

97. $4x + x + 6x = 11x$

98. $3x + 2y + 6x = 9x + 2y$

99. $14y - 3y - x = 11y - x$

100. $3xy + 5y + 2xy + 8y = 5xy + 13y$

101. $x + 2 = 9 \Rightarrow x = 7$

check: $7 + 2 \stackrel{?}{=} 9,\ 9 = 9$

102. $8 - n = 2 \Rightarrow n = 6$

check: $8 - 6 \stackrel{?}{=} 2,\ 2 = 2$

103.
$$\begin{aligned} (3 + x) + 1 &= 8 \\ (x + 3) + 1 &= 8 \\ x + (3 + 1) &= 8 \\ x + 4 &= 8 \\ x &= 4 \end{aligned}$$

104.
$$\begin{aligned} 2 + (n + 7) &= 10 \\ (n + 7) + 2 &= 10 \\ n + (7 + 2) &= 10 \\ n + 9 &= 10 \\ n &= 1 \end{aligned}$$

105. $9x = 27 \Rightarrow x = 3$

check: $9(3) \stackrel{?}{=} 27,\ 27 = 27$

106. $\frac{20}{x} = 5 \Rightarrow x = 4$

check: $\frac{20}{4} \stackrel{?}{=} 5,\ 5 = 5$

107. $12n - n = 22 \Rightarrow 11n = 22 \Rightarrow n = 2$

108. $y + 3y + 2y = 12 \Rightarrow 6y = 12 \Rightarrow y = 2$

109. (a) $18 - x = 3$ (b) $x = 15$

110. (a) $x+5=11$ (b) $x=6$

111. (a) $3(x\cdot 2)=12 \Rightarrow 6x=12 \Rightarrow$ (b) $x=2$

112. $30+30+90+160=\$310$

113. $3560-499-218-97=\$2746$

114. (a) $5021+759+2534+532$
$-799-533-88=\$7426$
(b) $7426\div 2=\$3713$

115. $(2(20)+2(25)+2(15)+2(18))(3)$
$=\$468$

Chapter 1 Test

1. $1525=1000+500+20+5$

2. (a) $7>2$ (b) $5>0$

3. (a) $2925=3000$ to nearest thousand
(b) $2925=2900$ to nearest hundred

4. (a) $3+(8+x)=3+(x+8)=(x+8)+3$
$=x+(8+3)=x+11$
(b) $5+y+2=y+5+2=y+7$
(c) $1+(n+2)+4=(n+2)+1+4$
$=(n+2)+5$
$=n+(2+5)=n+7$

5. $12{,}389+4+2302=14{,}695$

6. $244{,}869{,}201+19{,}077=244{,}888{,}278$

7. (a) $613-75=538$
check: $75+538=613$
(b)
$$\begin{array}{r} 20{,}105 \\ -\ 7{,}826 \\ \hline 12{,}279 \end{array}$$
check: $7826+12{,}279=20{,}105$

8. (a) $3xy+2y+4xy-2=7xy+2y-2$
(b) $2m+5+m+6mn=3m+5+6mn$

9. $2+1+7+5+(7+2)+6=30$ ft

10. $2(4)(y\cdot 2)=8(2y)=16y$

11. $3(y+4)=3y+3(4)=3y+12$

12. (a) $(432)(312)=134{,}784$
(b)
$$\begin{array}{r} 2031 \\ \times\ \ 129 \\ \hline 261{,}999 \end{array}$$

13. (a)
$$\begin{array}{r} 41 \\ 12\overline{)492} \\ \underline{48} \\ 12 \\ \underline{12} \\ 0 \end{array}$$
check: $12(41)=492$

(b) $5523\div 46=120$ R3
$$\begin{array}{r} 120 \\ 46\overline{)5523} \\ \underline{46} \\ 92 \\ \underline{92} \\ 3 \end{array}$$
check: $120(46)+3=5523$

14. (a) Seven subtracted from a number:
$n-7$
(b) The product of ten and a number:
$10n$

15. (a) $x-y\big|_{x=16,\,y=4}=16-4=12$
(b) $ab\big|_{a=9,\,b=7}=9(7)=63$

16. $6\cdot 6\cdot 6\cdot 6\cdot 6=6^5$

17. (a) $5^3 = 5 \cdot 5 \cdot 5 = 125$
(b) $10^5 = 10 \cdot 10 \cdot 10 \cdot 10 \cdot 10 = 100{,}000$

18. (a) *y* to the fourth power: y^4
(b) 7 cubed: 7^3

19. $24 \div 4 - 2 \cdot 3 = 6 - 6 = 0$

20. $6^2 - 7 + 3 \cdot 4 = 36 - 7 + 12 = 29 + 12 = 41$

21. $3 \cdot 2 + 4(7-1) = 6 + 4(6) = 6 + 24 = 30$

22. (a) $7 + x = 13 \Rightarrow x = 6$
check: $7 + 6 \stackrel{?}{=} 13,\ 13 = 13$
(b) $\frac{x}{4} = 2 \Rightarrow x = 8$
check: $\frac{8}{4} \stackrel{?}{=} 2,\ 2 = 2$
(c) $2(4x) = 72 \Rightarrow 8x = 72 \Rightarrow x = 9$
check: $2(4(9)) \stackrel{?}{=} 72,\ 72 = 72$
(d) $5 + (b+2) = 18 \Rightarrow (b+2) + 5 = 18$
$b + (2+5) = 18$
$b + 7 = 18$
$b = 11$
check: $5 + (11+2) \stackrel{?}{=} 18,\ 18 = 18$
(e) $9n - n = 32 \Rightarrow 8n = 32 \Rightarrow n = 4$
check: $9(4) - 4 \stackrel{?}{=} 32,\ 32 = 32$

23. $B - 155 = 275$

24. (a) $x \div 6 = 2$
(b) $x = 12$

25. (a) $x - 3 = 1$
(b) $x = 4$

26. (a) $25(412) + 18(280) = \$15{,}340$
(b) $P = 15{,}340 - 7350 = \$7990$

27. $4(3) = 12$ different sandwiches

28. $1540 - 265 - 78 - 57 = \$1140$

29. $2(525) + 200 + 40 = \$1290$

30. (a) $800 + 200 + 100 + 200 + 300 = \$1600$
(b) $1900 - 1600 = \$300$

31. $(5000 \div 2) \cdot 3 = 7500$ points

Chapter 2

Pretest Chapter 2

1. (a) $-13 < 13$ (b) $6 < -85$

2. The opposite of -18 is $\underline{18}$.

3. $-(-y)\big|_{y=-11} = -(-(-11)) = -(11) = -11$

4. (a) $|4| = 4$
(b) $|-7| = -(-7) = 7$

5. $-5 + (-2) = -7$

6. $(-1) + 8 + (-6) = 7 + (-6) = 1$

7. (a) $x + 8\big|_{x=-1} = -1 + 8 = 7$
(b) $x + 8\big|_{x=-9} = -9 + 8 = -1$

8. $3 - 5 = 3 + (-5) = -2$

9. $-8 - 2 = -8 + (-2) = -10$

10. $5 - (-14) + 2 = 5 + (-(-14) + 2$
$= 5 + 14 + 2$
$= 19 + 2 = 21$

11. $1200 - (-250) = 1200 + 250 = 1450$ ft

12. $(-5)(4) = -20$

13. $(-6)(-3)(-2)(-1) = 18(2) = 36$

14. (a) $(-4)^2 = (-4)(-4) = 16$
(b) $-4^2 = -4 \cdot 4 = -16$

15. (a) $12 \div (-3) = -4$
(b) $(-12) \div (-3) = 4$

16. $\dfrac{-42}{6} = -7$

17. $5 - 12(9 - 10) = 5 - 12(-1)$
$= 5 - (-12) = 5 + (-(-12))$
$= 5 + 12 = 17$

18. $7 - 24 \div 6(-2)^2 - 3 = 7 - 4(-2)^2 - 3$
$= 7 - 4(4) - 3$
$= 7 - 16 - 3$
$= -9 - 3$
$= -12$

19. $5x + (-3x) = 2x$

20. $-3x^2 + y\big|_{x=-1, y=-6} = -3(-1)^2 + (-6)$
$= -3(1) + (-6)$
$= -3 + (-6) = -9$

21. $-3(x - 1) = -3x - (-3)(1) = -3x + 3$

2.1 Exercises

1. $-(-1)$: the opposite of negative one

3. Negative four minus two: $-4 - 2$

5. $-5 < -2$

7. $9 > -3$

9. $-5 < 5$

11. $-291 < -5$

13. $-1250 < 5$

15. $298 > -3$

17. A represents the larger number.

19. $\underline{\ +\ }$ A tax increase of \$100

21. $\underline{\ -\ }$ A loss of \$100

23. $\underline{\ -\ }$ A discount of \$10

25. $\underline{\ +\ }$ A plane ascends 1000 ft

27.

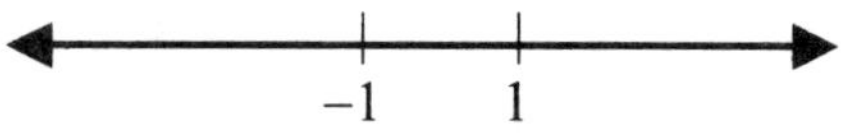

29. The opposite of –5 is $\underline{\ 5\ }$.

31. The opposite of 16 is $\underline{\ -16\ }$.

33. $-(-11) = \underline{\ 11\ }$

35. $-(5) = \underline{\ -5\ }$

37. $-(-(-7)) = -(7) = -7$

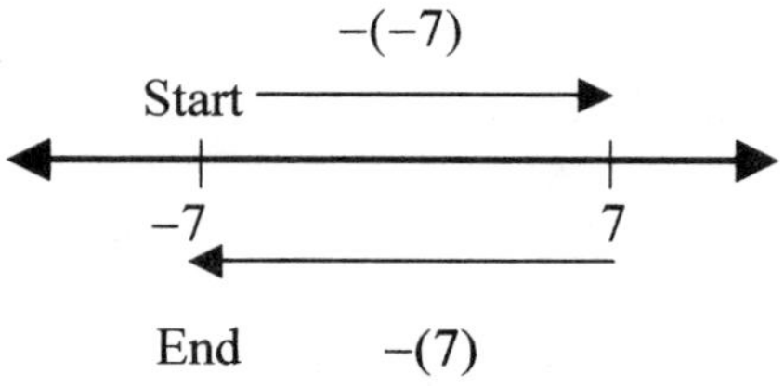

39. $-(-(-(-1)))) = -(-(1)) = 1$

41. $-(-(13)) = -(-13)$
$= 13$

43. $-(-a)\big|_{a=6} = -(-6)$
$= 6$

45. $-(-(-x)) = -(x) = -x\big|_{x=-1}$
$= -(-1) = 1$

47. $-(-(-(-y))) = -(-(y)) = -(-y)$
$= y\big|_{y=-2} = -2$

49. $|8| = 8$

51. $|-5| = -(-5) = 5$

53. $|-16| = -(-16) = 16$

55. $|44| = 44$

57. $|-3| = 3,\ |1| = 1,\ 3 > 1 \Rightarrow |-3| > |1|$

59. $|8| = 8,\ |-8| = 8, \Rightarrow |8| = |-8|$

61. $|16| = 16,\ |-9| = 9 \Rightarrow |16| > |-9|$

63. $|-35| = 35,\ |-8| = 8 \Rightarrow |-35| > |-8|$

65. $-|-3| = -(-(-3)) = -(3) = -3$

67. $-|14| = -14$

69. $|-33| = 33,\ |12| = 12 \Rightarrow |-33| > |12|$
-33 has the larger absolute value.

71. $|129| = 129,\ |-112| = 112 \Rightarrow |129| > |-112|$
129 has the larger absolute value.

73. (a) Fargo
(b) Positive: Boston and Cleveland
Negative: Albany, Anchorage, Fargo

Cumulative Review Problems

75. $5009 - 258 = 4751$

77. $(256)(91) = 23,296$

79. $2600 - 400 - 1200 - 350 = \$570$

2.2 Exercises

1. The sum of two negative numbers is negative.

3. $-10\ °F + (-5\ °F) = -15\ °F$

5. $\$100 + \$50 = \$150$

7.(a)

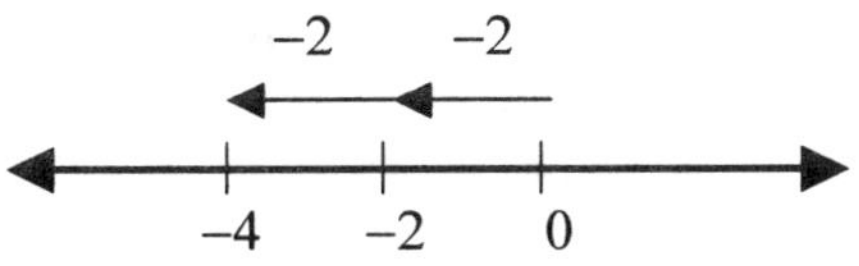

(b) Negative
(c) $-2 + (-2)$
(d) From number line, sum is -4

9. (a) $-11 + (-13) = -24$
(b) $11 + 13 = 24$

11. (a) $-29 + (-39) = -68$
(b) $29 + 39 = 68$

13. (a) $-43 + (-18) = -61$
(b) $43 + 18 = 61$

15. $+300\text{ ft} + (-400\text{ ft}) = -100\text{ ft}$

17. $-\$400 + (+\$500) = \$100$

19. (a) $-3 + 10$ (b) Positive
(c) $-3 + 10 = 7$

21. (a) $2 + (-4)$ (b) Negative
(c) $2 + (-4) = -2$

23. (a) $6 + (-8) = -2$ (b) $-6 + 8 = 2$

25. (a) $5 + (-1) = 4$ (b) $-5 + 1 = -4$

27. (a) $22 + (-16) = 6$ (b) $-22 + 16 = -6$

29. $12 + (-11) = 1$

31. $-9 + 6 = -3$

33. $-7 + 14 = 7$

35. $22 + (-10) = 12$

37. $-33 + (-5) = -38$

39. $-27 + (-12) = -27$

41. (a) $3 + (-1) = 2$ (b) $-3 + (-1) = -4$

43. (a) $-9 + (-11) = -20$
(b) $9 + (-11) = -2$

45. $2 + (-2) = 0$

47. $-8 + 8 = 0$

49. $-350 + 350 = 0$

51. $452 + (-452) = 0$

53. $6 + (-9) + 1 + (-3) = -3 + (-2) = -5$

55. $-21 + 16 + (-33) = -5 + (-33) = -38$

57. $17 + (-29) + (-34) + 43 = -12 + 9 = -3$

59. (a) $x + 3\big|_{x=-2} = -2 + 3 = 1$
(b) $x + 3\big|_{x=-7} = -7 + 3 = -4$

61. (a) $x+(-8)\big|_{x=4} = 4+(-8) = -4$
(b) $x+(-8)\big|_{x=-5} = -5+(-8) = -13$

63. (a) $-2+x+y\big|_{x=-6,y=4} = -2+(-6)+4 = -4$
(b) $-2+x+y\big|_{x=5,y=-9} = -2+5+(-9) = -6$

65. $30,000+(-20,000) = \$10,000$

67. $-121+200 = \$79$

69. $-75+(-150) = -225$ ft

71. $5421+(-1256)+8001+(-3,444,589)$
$= -3,432,423$

73. $3+(-33) = -30$

Cumulative Review Problems

75. $2(3x) = 6x$

77. $3(x-4) = 3x-3(4) = 3x-12$

79. $12+(-4+8)+11+(-7+15) = 35$ people

2.3 Exercises

1. To subtract two numbers, we change the subtraction sign to <u>addition</u> and take the <u>opposite</u> of the second number. Then we <u>add</u>.

3. (a) $-3+6 = n$ (b) $n = 3$

5. $\$20-\$30 = \$20+(-\$30) = -\$10$

7. $\$7-\$8 = \$7+(-\$8) = -\$1$

9. $10-11 = 10+(-11) = -1$

11. $9-10 = 9+(-10) = -1$

13. (a) $7-4 = 7+(-4) = 3$
(b) $15-7 = 15+(-7) = 8$
(c) $10-8 = 10+(-8) = 2$

15. $-6-4 = -6+(-4) = -10$

17. $-8-(-6) = -8+(-(-6)) = -8+6 = -2$

19. $-5-4 = -5+(-4) = -9$

21. $-8-(-8) = -8+(-(-8)) = -8+8 = 0$

23. $2-7 = 2+(-7) = -5$

25. $4-(-2) = 4+(-(-2)) = 4+2 = 6$

27. $3-7 = 3+(-7) = -4$

29. $5-(-9) = 5+(-(-9)) = 5+9 = 14$

31. $5-19 = 5+(-19) = -14$

33. $-7-(-8) = -7+(-(-8)) = -7+8 = 1$

35. $-13-18 = -13+(-18) = -31$

37. $52-(-1) = 52+(-(-1)) = 52+1 = 53$

39. $80-90 = 80+(-90) = -10$

41. $-85-(-20) = -85+(-(-20))$
$= -85+20 = -65$

43. $7-9-3-8 = -2-3-8 = -5-8 = -13$

45. $8-1-9-5 = 7-9-5 = -2-5 = -7$

47. $9-10-2+3 = -1-2+3 = -3+3 = 0$

49. $5+8-6-4=13-6-4=7-4=3$

51. $2-1-9-7=1-9-7=-8-7=-15$

53. $-6-(-3)+(-7)=-3+(-7)=-10$

55. $-7-(-2)-(-5)=-5-(-5)=0$

57. $-3-(-8)-(-6)=5-(-6)=11$

59. $3556-(-150)=3706$ ft

61. $326-(-18)=344$ ft

63. (a) Brownsville, Texas
(b) $77\text{ °F}-(-9\text{ °F})=86\text{ °F}$

65. $-11-(-15)=4$ points

67. $-15<-13 \rightarrow$ Woods is the winner.

69. $x-15\big|_{x=-9}=-9-15=-24$

71. $x-10\big|_{x=-1}=-1-10=-11$

73. $19-y\big|_{y=-6}=19-(-6)=19+6=25$

75. $14-y\big|_{y=-2}=14-(-2)=14+2=16$

77. $-7-x\big|_{x=-3}=-7-(-3)=-7+3=-4$

79. $-2-x\big|_{x=-5}=-2-(-5)=-2+5=3$

81. $-3009-893=-3000+(-893)$
$=-3902$

83. $-2001-(-987)=-2001+(-(-987))$
$=-2001+987=-1014$

Cumulative Review Problems

85. $12-3(4-1)=12-3(3)=12-9=3$

87. $3+[3+2(8-6)]=3+[3+2(2)]$
$=3+[3+4]=3+7=10$

89. $8670 \div 85=102$ min $=1$ hr 42 min

2.4 Exercises

1. If you multiply 4 negative numbers, the product will be a <u>positive</u> number.

3. The quotient of a positive number and a <u>negative</u> number is negative.

5. $3(-4)=(-4)+(-4)+(-4)=-12$

7. $4(-6)=(-6)+(-6)+(-6)+(-6)=-24$

9. $2(-3)=(-3)+(-3)=-6$

11. $(-2)(-9)=18$

13. $-3(-6)=18$

15. $5(-6)=-30$

17. $-8(3)=-24$

19. (a) $4(2)=8$ (b) $4(-2)=-8$
(c) $-4(2)=-8$ (d) $-4(-2)=8$

21. (a) $5(2)=10$ (b) $-5(-2)=10$
(c) $-5(2)=-10$ (d) $5(-2)=-10$

23. $(-1)(3)(-236)(42)(-16)(-90)$ is a positive product because it contains an even number of negative factors.

25. $(-943)(-721)(-816)(-96)(-51)$ is a negative product because it contains an odd number of negative factors.

27. $(-3)(-2)(-3)(-4) = (6)(12) = 72$

29. $3(-7)(-2) = (-21)(-2) = 42$

31. $3(-1)(5)(-6) = (-3)(-30) = 90$

33. $(-2)(-1)(4)(-5) = 2(-20) = -40$

35. $(-5)(4)(-3)(2)(-1) = (-20)(-6)(-1)$
$= (120)(-1) = -120$

37. The value of $(-2)^{13}$ is a negative number because it contains an odd number of factors.

39. The value of $(-96)^{52}$ is a positive number because it contains an even number of factors.

41. $(-5)^2 = (-5)(-5) = 25$

43. $(-5)^3 = (-5)(-5)(-5) = -125$

45. (a) $(-3)^2 = (-3)(-3) = 9$
(b) $(-3)^3 = (-3)(-3)(-3) = -27$

47. (a) $(-1)^{11} = -1$, 11 is odd
(b) $(-1)^{24} = 1$, 24 is even

49. (a) $-4^2 = -(4)(4) = -16$
(b) $(-4)^2 = (-4)(-4) = 16$

51. (a) $-2^3 = -(2)(2)(2) = -8$
(b) $(-2)^3 = (-2)(-2)(-2) = -8$

53. $20 \div (-5) = -4$

55. $\dfrac{-36}{6} = -6$

57. $-16 \div (-8) = 2$

59. $\dfrac{-49}{-7} = 7$

61. (a) $35 \div 7 = 5$ (b) $35 \div (-7) = -5$
(c) $-35 \div 7 = -5$ (d) $-35 \div -7 = 5$

63. (a) $40 \div 8 = 5$ (b) $40 \div (-8) = -5$
(c) $-40 \div 8 = -5$ (d) $-40 \div (-8) = 5$

65. (a) $22 \div (-2) = -11$
(b) $22(-2) = -44$

67. (a) $-4 \div (-2) = 2$ (b) $-4(-2) = 8$

69. $350(-2) = -\$700$

71. $-30(3) = -90$ m or 90 m left of 0

73. $(-578)(-698) = 403,444$

75. $\dfrac{-1357}{23} = -59$

77. $\dfrac{x}{-3} = 8 \Rightarrow x = -24$

79. $x^2\big|_{x=-1} = (-1)^2 = (-1)(-1) = 1$

81. $\dfrac{-x}{y}\Bigg|_{x=36,\,y=6} = \dfrac{-36}{6} = -6$

Cumulative Review Problems

83. $2^2+3(5)-1=4+15-1=19-1=18$

85. $2^3+(4\div 2+6)=8+(2+6)=8+8=16$

87. $3261\div 1087=3$ sec

2.5 Exercises

1. $-4(0)=0$ and 0 is neither positive or negative.

3. $-2+3.5=1.5$

5. $5+7(2-6)=5+7(-4)=5+(-28)=-23$

7. $5(-3)(4-7)+9=-15(-3)+9$
$=45+9$
$=54$

9. $-3(6\div 3)+7=-3(2)+7=-6+7=1$

11. $3(-2)(9-5)-10=-6(4)-10$
$=-24-10$
$=-34$

13. $-24\div 12-8=-2-8=-10$

15. $(-3)^2+6(-9)=9+(-54)=-45$

17. $(-2)^3-7(8)=-8-56=-64$

19. $36\div(-6)+(-6)=-6+(-6)=-12$

21. $(-2)^3+3(-8)=-8+(-24)=-32$

23. $12-20\div 4(-4)^2+9$
$=12-20\div 4(16)+9=12-5(16)+9$
$=12-80+9=-68+9=-59$

25. $\dfrac{[-50\div 2+3]}{[20-9]}=\dfrac{-25+3}{11}=\dfrac{-22}{11}=-2$

27. $\dfrac{[3^2+4(-6)]}{[-3+(-2)]}=\dfrac{9+(-24)}{-5}=\dfrac{-15}{-5}=3$

29. $\dfrac{[(30-15\div 3)+(-5)]}{[5-10]}=\dfrac{(30-5)+(-5)}{-5}$
$=\dfrac{25+(-5)}{-5}$
$=\dfrac{20}{-5}=-4$

31. $-16\div[-4\cdot(8\div(-2))]$
$=-16\div[-4\cdot(-4)]=-16\div 16=-1$

33. $[(3+24)\div(-3)]\cdot[2+(-3)^2]$
$=[27\div(-3)]\cdot[2+9]=-9\cdot 11=-99$

35. $30,000+3(-2000)+1000$
$=30,000+(-6000)+1000$
$=24,000+1000=25,000$ ft

37. $14(-3)+9(1)=-42+9=-33$

39. $7(+3)+5(-1)+4(+1)=21-5+4=+20$

41. $1(10)+1(5)+1(3)+(1(-1)+1(-1))+2(-1)$
$=10+5+3+(-1-1)-2$
$=15+3-2-2=18-4=14$ points

43. $1(20)+1(5)+2(3)+(2(-1)+2(-1))$
$+2(-1)=20+5+6+(-2-2)-2$
$=25+6-4-2=31-6$
$=25$ points

45. $(-9)^2 - 17(-85) = 81 - (-1445)$
$= 81 + 1445$
$= 1526$

47. $-65 - 144 \div 9(-14) + 64$
$= -65 - 16(-14) + 64$
$= -65 - (-224) + 64 = -65 + 224 + 64$
$= 159 + 64 = 223$

49. $3 + x - 2(-4) = 7 - (-13)$
$x + 3 - (-8) = 7 + 13$
$x + 3 + 8 = 20$
$x + 11 = 20$
$x = 9$

Cumulative Review Problems

51. $2(x + 3) = 2x + 2(3) = 2x + 6$

53. $3(x - 2) = 3x - 3(2) = 3x - 6$

55. $(5 \cdot 32) \div 20 = 160 \div 20 = 8$ bills

2.6 Exercises

1. $-8x + 3x = (-8 + 3)x = -5x$

3. $2x + (-3x) = (2 + (-3))x = -x$

5. $-5x - 7x = (-5 - 7)x = -12x$

7. $-7a - (-2a) = (-7 - (-2))a = -5a$

9. $9x + (-7x) = (9 + (-7))x = 2x$

11. $-7x + (-6x) = (-7 + (-6))x = -13x$

13. $7a + (-9b) = 7a - 9b$

15. $-4m + (-8n) = -4m - 8n$

17. $3x + (-y) = 3x - y$

19. $2a - (-3b) = 2a + 3b$

21. $-8y + 5x + 2y = -8y + 2y + 5x$
$= -6y + 5x$

23. $6x + 2y + (-8x) = 6x + (-8x) + 2y$
$= -2x + 2y$

25. $9x + 3y + (-5x) = 9x + (-5x) + 3y$
$= 4x + 3y$

27. $-8x - 4x - y = (-8 - 4)x - y = -12x - y$

29. $3a + 2x - 5a + 7ax - x$
$= 3a - 5a + 2x - x + 7ax$
$= -2a + x + 7ax$

31. $6x + 5y - 10x - 2y$
$= 6x - 10x + 5y - 2y$
$= -4x + 3y$

33. $4x + 2y - 6x - 7 = 4x - 6x + 2y - 7$
$= -2x + 2y - 7$

35. $4 + 3ab - 2 - 9ab = 4 - 2 + 3ab - 9ab$
$= 2 - 6ab$

37. $5 + 5xy - 9 - xy = 5 - 9 + 5xy - xy$
$= -4 + 4xy$

39. $8 - 2ab - 9 - 7ab = 8 - 9 - 2ab - 7ab$
$= -1 - 9ab$

41. (a) $2 - 7 + 3 = -5 + 3 = -2$
(b) $2x - 7x + 3x = -5x + 3x = -2x$

43. (a) $3-8+4=-5+4=-1$
(b) $3x-8x+4x=-5x+4x$
$=-1x$
$=-x$

45. (a) $2-6+1=-4+1=-3$
(b) $2x-6x+1x=-4x+x=-3x$

47. $x+3y\big|_{x=-3,y=-2}=-3+3(-2)=-3-6=-9$

49. $m-6n\big|_{m=6,n=-3}=6-6(-3)=6+18=24$

51. $ab-6\big|_{a=-1,b=5}=(-1)(5)-6=-5-6=-11$

53. $\left.\frac{x}{y}\right|_{x=-25,y=5}=\frac{-25}{5}=-5$

55. $\left.\frac{2y}{z}\right|_{y=-4,z=-2}=\frac{2(-4)}{-2}=\frac{-8}{-2}=4$

57. $\left.\frac{(a+b)}{4}\right|_{a=-9,b=5}=\frac{-9+5}{4}=\frac{-4}{4}=-1$

59. $9t^2\big|_{t=-3}=9(-3)^2=9(9)=81$

61. $8x-x^2\big|_{x=-5}=8(-5)-(-5)^2$
$=-40-25=-65$

63. $\left.\frac{(x^2-x)}{2}\right|_{x=-4}=\frac{(-4)^2-(-4)}{2}$
$=\frac{16+4}{2}=\frac{20}{2}=10$

65. $-3(x+2)=-3x+(-3)(2)$
$=-3x-6$

67. $-9(y-1)=-9y-(-9)(1)=-9y+9$

69. $-2(m-2)=-2m-(-2)(2)=-2m+4$

71. $-1(x+5)=-1x-5=-x-5$

73. $4(-2+y)=4(-2)+4y=-8+4y$

75. $2(-4+a)=2(-4)+2a=-8+2a$

77. $s=v-32t=-8-32(4)=-136$ ft/sec

79. $C=\frac{5F-160}{9}=\frac{5(5)-160}{9}=-15$ °C

81. $x^3-y^4\big|_{x=-9,y=2}=(-9)^3-2^4=-729-16$
$=-745$

83. $x^{212}+x^{305}\big|_{x=-1}=(-1)^{212}+(-1)^{305}$
$=1+(-1)=0$

85. $\left.\frac{x^3}{2}\right|_{x=0}=\frac{0^3}{2}=\frac{0}{2}=0,$ yes

87. True. If a number is positive, its opposite is negative and the product is negative. If a number is negative its opposite is positive and the product is negative.

Cumulative Review Problems

89. $24,813=24,800$ to nearest hundred

91. $P=2l+2w=2(6)+2(3)$
$P=18$ ft

93. $5,580,000\div 30=186,000$ mi in one sec
$5,580,000(2)$
$=11,160,000$ mi in one min

Putting Your Skills to Work, Problems for Individual Investigation

1. $|0-(-3)|=3$ hr earlier
10 A.M. − 3 hr gives 7 A.M. in Montevideo, Uruguay.

2. $|0-9|=9$ hr later
2 P.M. + 9 hr gives 11 P.M. in Seoul, South Korea.

3. $|3-(-7)|=10$ hr earlier
3 P.M. − 10 hr gives 5 A.M. in Denver, Colorado.

4. $|8-5|=3$ hr earlier
6 A.M. − 3 hr gives 3 A.M. in Karachi, Pakistan.

Putting Your Skills to Work, Problems for Group Investigation and Cooperative Study

1. $|10-(-6)|=16$ hr earlier
9 A.M. Monday − 16 hr gives 5 P.M on Sunday in Mexico City, Mexico.

2. 2 P.M. Tuesday − 16 hr gives 10 P.M. on Monday in Mexico City, so, yes, Victor can watch since he got home at 6 P.M.

3. Since the call takes 30 min it must be made by 5 P.M. in Bucharest which is noon in Montevideo and 10 A.M. in Boston. Since the earliest the call can be made is 8 A.M. in Boston which is 10 A.M. in Montevideo and 1 P.M. in Bucharest, the call must be made between 8 A.M. and 10 A.M. from the Boston office.

4. Since the call takes 30 min it must be made by 4 P.M. in Kurachi which is noon in Algiers and 8 A.M. in Montvideo. The engineer in Algiers must make the call at noon.

Chapter 2 Review

1. $-3<-1$

2. $-7<-2$

3. $4>-4$

4. $5>-5$

5. __+__ A profit of \$100

6. __−__ A drop in temperature of 16°.

7. The opposite of −16 is _16_.

8. $-(-(-8))=-(8)=-8$

9. $|5|=5$

10. $-|-11|=-(-(-11))=-(11)=-11$

11. $|-23|=23,\ |15|=15,\ 23>15\Rightarrow -23$ has a larger absolute value than 15.

12. $|-10|=10,\ |12|=12,\ 12>10\Rightarrow 12$ has a larger absolute value than −10.

13. (a) Justin made the most money in May.
(b) Justin lost the most money in March.

14. (a) Justin had a net gain in January, February, and May.
(b) Justin had a net loss in March and April.

15. $-\$500+(-\$200)=-\$700$

16. $\$100+\$200=\$300$

17. (a) $-43+(-16)=-59$
(b) $43+16=59$

18. (a) $-27+(-39)=-66$
(b) $27+39=66$

19. $-\$25,000+\$15,000=-\$10,000$
The company had a net loss.

20. $-\$14+\$25=\$11$, Terry had a net profit.

21. (a) $-32\ °F+20\ °F$
(b) Negative
(c) $-32\ °F+20\ °F=-12\ °F$

22. (a) $2+(-8)=-6$
(b) $-2+8=6$
(c) $-2+(-8)=-10$

23. (a) $32+(-18)=14$
(b) $-32+18=-14$
(c) $-32+(-18)=-50$

24. $3+(-5)+8+(-2)=-2+6=4$

25. $24+(-52)+(-12)+(-56)$
$=-28+(-68)=-96$

26. $x+6|_{x=-1}=-1+6=5$

27. $-2+x+y|_{x=-3,y=-11}=-2+(-3)+(-11)$
$=-5+(-11)=-16$

28. $-900+(220)=-680$ ft

29. $(-240)+350+400+(-800)=-290$ ft

30. $-7-5=-7+(-5)=-12$

31. $-9-(-4)=-9+4=-5$

32. $-5-3=-5+(-3)=-8$

33. $-6-(-2)=-6+2=-4$

34. $-3-8+6=-11+6=-5$

35. $6-(-4)+(-5)=6+4+(-5)$
$=10+(-5)=5$

36. $-4-(-2)=-4+2=-2$

37. $6-9-2-8=-3-2-8=-5-8=-13$

38. $-6-(-9)+(-1)=-6+9+(-1)$
$=3+(-1)=2$

39. $30,000-(-20,000)=\$50,000$

40. $10,000-(-30,000)=\$40,000$

41. (a) $6(3)=18$
(b) $6(-3)=-18$
(c) $-6(3)=-18$
(d) $-6(-3)=18$

42. (a) $5(2)=10$
(b) $5(-2)=-10$
(c) $-5(2)=-10$
(d) $-5(-2)=10$

43. $-7(-5)=35$

44. $-2(5)=-10$

45. $3(-4)=-12$

46. $-4(-1)=4$

47. $(-2)(-5)(-9)=10(-9)=-90$

48. $(-2)(-8)(-1)(-4)=16(4)=64$

49. $(-5)(1)(-2)(4)(-6)=(-5)(-8)(-6)$
$=40(-6)=-240$

50. $(-5)^3=(-5)(-5)(-5)=-125$

51. $-2^2=-(2)(2)=4$

52. $(-2)^2=(-2)(-2)=4$

53. (a) $49\div7=7$
(b) $49\div(-7)=-7$

54. (a) $-30\div5=-6$
(b) $-30\div(-5)=6$

55. (a) $-44\div(-4)=11$
(b) $9(-5)=-45$
(c) $(-11)(-3)=33$
(d) $\frac{25}{-5}=-5$

56. (a) $12\div(-4)=-3$
(b) $5(-8)=-40$
(c) $-12(-2)=24$
(d) $\frac{45}{-9}=-5$

57. $4-1(6-9)=4-(6-9)=4-6+9$
$=-2+9=7$

58. $3(-5)(2-6)+8=-15(-4)+8$
$=60+8=68$

59. $-2^2+3(-4)=-4+(-12)=-16$

60. $\frac{[-32\div8+4]}{[7-9]}=\frac{-4+4}{-2}=\frac{0}{-2}=0$

61. $12+3(-5)+(-2)=12+(-15)+(-2)$
$=-3+(-2)=-5$ °F

62. $-5y+3y=(-5+3)y=-2y$

63. $-4y+3x+9y=(-4+9)y+3x$
$=5y+3x=3x+5y$

64. $-3a-6a=(-3-6)a=-9a$

65. $7x+9y-6x-11y=(7-6)x+(9-11)y$
$=x+(-2)y$
$=x-2y$

66. $3+5z-7+2yz-8z$
$=3-7+(5-8)z+2yz=-4-3z+2yz$

67. $-8+7y+7-2y=-8+7+(7-2)y$
$=-1+5y$

68. $a+3b\big|_{a=-1,b=-4}=-1+3(-4)$
$=-1-12=-13$

69. $2x-y\big|_{x=-2,y=-1}=2(-2)-(-1)$
$=-4+1=-3$

70. $\frac{x+y}{4}\bigg|_{x=-1,y=-7}=\frac{-1+(-7)}{4}=\frac{-8}{4}=-2$

71. $a^2-b\big|_{a=-3,b=9}=(-3)^2-9=9-9=0$

72. $-6(x+2)=-6x+(-6)(2)=-6x-12$

73. $-2(a-1)=-2a-(-2)(1)=-2a+2$

74. $4(-2+x)=4(-2)+4x=-8+4x$

Chapter 2 Test

1. $-234<-5$

2. $|4|=4,\ |-18|=18,\ 4<18 \Rightarrow |4|<|-18|$

3. ___–___ 14 points

4. $-(-(-2))=-(2)=-2$

5. The opposite or 10 is $\underline{-10}$.

6. (a) $|12|=12$ (b) $|-3|=-(-3)=3$

7. $-6+8=2$

8. $-6+(-4)=-10$

9. $-20+5+(-1)+(-3)=-15+(-1)+(-3)$
$=-16+(-3)$
$=-19$

10. (a) $-10\ °F+15\ °F$
(b) $-10\ °F+15\ °F=5\ °F$

11. $20{,}000+(-5000)=\$15{,}000$ profit

12. $3700-(-592)=4292$ ft

13. $12-18=12+(-18)$
$=-6$

14. $-1-11=-1+(-11)=-12$

15. $3-(-10)=3+10=13$

16. $-14-3+(-6)-1=-17+(-6)-1$
$=-23-1$
$=-24$

17. (a) $-7+x+y\big|_{x=-6,\,y=-3}=-7+(-6)+(-3)$
$=-13+(-3)=-16$
(b) $-7+x+y\big|_{x=-7,\,y=6}=-7+(-7)+6$
$=-14+6=-8$

18. $(7)(-3)=-21$

19. $(-8)(-4)=32$

20. $(-5)(-2)(-1)(3)=10(-3)=-30$

21. (a) $(-5)^2=(-5)(-5)=25$
(b) $(-5)^3=(-5)(-5)(-5)=-125$
(c) $-5^2=-(5)(5)=25$

22. (a) $-8\div 2=-4$ (b) $-8\div(-2)=4$

23. $\dfrac{-22}{11}=-2$

24. $7+5x+2xy-8x-6xy$
$=7+(5-8)x+(2-6)xy$
$=7+(-3)x+(-4)xy=7-3x-4xy$

25. $-4(11-13)+5=-4(-2)+5=8+5=13$

26. $2-35\div 5(-3)^2-6=2-7(-3)^2-6$
$=2-7(9)-6$
$=2-63-6$
$=-61-6=-67$

27. $\dfrac{[-8+2(-3)]}{[14-21]}=\dfrac{-8+(-6)}{-7}=\dfrac{-14}{-7}=2$

28. $IQ = \frac{100M}{C} = \frac{100(23)}{20} = 115$

29. $\left.\frac{2x+y^2}{-7}\right|_{x=-1,y=-4} = \frac{2(-1)+(-4)^2}{-7}$

$= \frac{-2+16}{-7} = \frac{14}{-7} = -2$

30. $-6(a+7) = -6a + (-6)(7) = -6a - 42$

31. $-2(x-1) = -2x - (-2)(1) = -2x + 2$

Cumulative Test for Chapters 1 and 2

1. 5280 = 5300 to nearest hundred

2. $1650 - 1475 = 175$ years

3. (a) $167,350 - 86,000 = \$81,350$
(b) $81,350 \div 2 = \$40,675$

4. $6(7x) = 42x$

5. $3(y \cdot 8) = 3(8y) = 24y$

6. $400(7n) = 2800n$

7. $(219)(67) = 14,673$

8. $2172 \div 14 = 155$ R2

```
    155
14)2172
   14
    77
    70
     72
     70
      2
```

9. $3 + x|_{x=10} = 3 + 10 = 13$

10. $\frac{x}{2} = 4 \Rightarrow x = 8$, check: $\frac{8}{2} \stackrel{?}{=} 4$, $4 = 4$

11. Double some number equals 28:
$2x = 28$

12. $5 + x = 8 \Rightarrow x = 3$

13. $-8 > -617$

14. $|-17| = 17$, $|-2| = 2$, $17 > 2 \Rightarrow |-17| > |-2|$

15. $-(-(-(-6))) = -(-(6))$
$= -(-6)$
$= 6$

16. $-|-1| = -(-(-1))$
$= -(1)$
$= -1$

17. $5 + (-6) = -1$

18. $-10 - 8 = -10 + (-8)$
$= -18$

19. $-7 - 6 - 4 - 8 = -13 - 4 - 8$
$= -17 - 8$
$= -25$

20. $(-18) \div (-9) = 2$

21. $(-2)^5 = (-2)(-2)(-2)(-2)(-2)$
$= -32$

22. $(-2)(-1)(4)(3)(-2) = 2(12)(-2)$
$= -48$

23. $-4+15\div 5(-3)^2-1=-4+3(9)-1$
$=-4+27-1$
$=23-1$
$=22$

24. $3mn-7mn+4m=(3-7)mn+4m$
$=-4mn+4m$

25. $x-2y^3+1\Big|_{x=-1,\,y=-3}=-1-2(-3)^3+1$
$=-1-2(-27)+1$
$=-1-(-54)+1$
$=-1+54+1=54$

Chapter 3

Pretest Chapter 3

1. $x+17=13 \Rightarrow x+17-17=13-17$
$x+0=-4 \Rightarrow x=-4$

check: $-4+17\overset{?}{=}13,\ 13=13$

2. $-12=y-9 \Rightarrow -12+9=y-9+9$
$-3=y+0 \Rightarrow -3=y \Rightarrow y=-3$

check: $-12\overset{?}{=}-3-9,\ -12=-12$

3. $12-15=y-11 \Rightarrow y-11=-3$
$y-11+11=-3+11 \Rightarrow y+0=8$

$y=8$, check: $12-15\overset{?}{=}8-11,\ -3=-3$

4. $\angle a+\angle b=180° \Rightarrow \angle a+29°=180°$
$\angle a=180°-29°=151°$

5. (a) $\angle a=\angle b-40°$
(b) $60°=\angle b-40°$
$60°+40°=\angle b-40°+40°=\angle b+0°$
$\angle b=100°$

6. $-7y=56 \Rightarrow \dfrac{-7y}{-7}=\dfrac{56}{-7} \Rightarrow y=-8$

check: $-7(-8)\overset{?}{=}56,\ 56=56$

7. $5(2x)=\dfrac{60}{3}=20 \Rightarrow 10x=20 \Rightarrow x=2$

check: $5(2\cdot 2)\overset{?}{=}\dfrac{60}{3},\ 20=20$

8. $J=2B$

9. (a) $A=3R$
(b) $3=3R \Rightarrow R=\dfrac{3}{3}=1$

10. $P=4s=20 \Rightarrow s=\dfrac{20}{4}=5$ in.

11. $P=2L+2W=2(5W)+2W=12W=60$

$W=\dfrac{60}{12}=5$ ft

12. $A=bh=12(8)=96$ ft^2

13. $A=s^2=11^2=121$ ft^2

14. $A=LW \Rightarrow 88=11x \Rightarrow x=\dfrac{88}{11}=8$ yd

15. $A=bh \Rightarrow 108=9h \Rightarrow h=\dfrac{108}{9}=12$ m

16. $V=LWH \Rightarrow 60=3(5)H$

$H=\dfrac{60}{15}=4$ in.

17. $y\cdot y^4=y^{1+4}=y^5$

18. $x^5\cdot x^3=x^{5+3}=x^8$

19. $2^6\cdot 2^4=2^{6+4}=2^{10}$

20. $x^2y^4=x^2y^4$

21. $(5x^5)(3x^6)=15x^{5+6}=15x^{11}$

22. $(-2y)(y^4)=-2y^{1+4}=-2y^5$

23. $2y^3(3y-6)=6y^4-12y^3$

24. $(2x^2+3x)(-4x)=-8x^3-12x^2$

25. $A=LW=(2x^4-1)x=2x^5-x$

3.1 Exercises

1. The sum of two opposite numbers is equal to <u>zero</u>.

3. To solve $x-6=2$, we <u>add</u> 6 to both sides of the equation.

5. $4+\underline{(-4)}=0$

7. $-9+\underline{(9)}=0$

9. $17+\underline{(-17)}=0$

11. $-28+\underline{(28)}=0$

13. $x+8+\underline{(-8)}=x$

15. $m-2+\underline{(2)}=m$

17. $y-7+\underline{(7)}=y$

19. $x+9+\underline{(-9)}=x$

21. $y-15=5 \Rightarrow y-15+15=5+15$
$y+0=20 \Rightarrow y=20$

check: $20-15\stackrel{?}{=}5,\ 5=5$

23. $13+x=22 \Rightarrow 13-13+x=22-13$
$0+x=9 \Rightarrow x=9$

check: $13+9\stackrel{?}{=}22,\ 22=22$

25. $n-43=-74 \Rightarrow n-43+43=-74+43$
$n+0=-31 \Rightarrow n=-31$

check: $-31-43\stackrel{?}{=}-74,\ -74=-74$

27. $y+44=-50 \Rightarrow y+44-44=-50-44$
$y+0=-94 \Rightarrow y=-94$

check: $-94+44\stackrel{?}{=}-50,\ -50=-50$

29. $1=x-13 \Rightarrow 1+13=x-13+13$
$14=x+0 \Rightarrow x=14$

check: $1\stackrel{?}{=}14-13,\ 1=1$

31. $29=y+11 \Rightarrow 29-11=y+11-11$
$18=y+0 \Rightarrow y=18$

check: $29\stackrel{?}{=}18+11,\ 29=29$

33. $-13=x+1 \Rightarrow -13-1=x+1-1$
$-14=x+0 \Rightarrow x=-14$

check: $-13\stackrel{?}{=}-14+1,\ -13=-13$

35. $4x-3x-3=8$
$x-3+3=8+3$
$x+0=11$
$x=11$

37. $6=2y-y+4$
$6=y+4 \Rightarrow 6-4=y+4-4$
$2=y+0 \Rightarrow y=2$

39. $3y-2y+3=1 \Rightarrow y+3=1$
$y+3-3=1-3 \Rightarrow y+0=-2$
$y=-2$

41. $-23+8+x=-2+13 \Rightarrow -15+x=11$
$-15+x+15=11+15=26$
$x+0=26 \Rightarrow x=26$

43. $4-9=a-1+14 \Rightarrow -5=a+13$
$-5-13=a+13-13=a+0 \Rightarrow a=-18$

45. $-45+9+m=-6+18 \Rightarrow -36+m=12$
$-36+36+m=+12+36 \Rightarrow 0+m=48$
$m=48$

47. $-1+4+x=-5+9 \Rightarrow 3+x=4$
$3-3+x=4-3 \Rightarrow 0+x=1 \Rightarrow x=1$

49. $3(7-11)=y-5 \Rightarrow 3(-4)=y-5$
$-12+5=y-5+5 \Rightarrow -7=y+0$
$y=-7$

51. (a) $x-8=22 \Rightarrow x-8+8=22+8$
$x=30$
(b) $x+8=22 \Rightarrow x+8-8=22-8$
$x=14$

53. (a) $x+6-6=-11-6 \Rightarrow x=-17$
(b) $x-6+6=-11+6 \Rightarrow x=-5$

55. (a) $-18-2=x+2-2 \Rightarrow x=-20$
(b) $-18+2=x-2+2 \Rightarrow x=-16$

57. $\angle a+\angle b=180° \Rightarrow \angle a+86°=180°$
$\angle a=180°-86° \Rightarrow \angle a=94°$

59. $\angle x+\angle y=180° \Rightarrow \angle x+112°=180°$
$\angle x=180°-112° \Rightarrow \angle x=68°$

61. (a) $\angle x=\angle y+30°$
(b) $\angle x=\angle y+30° \Rightarrow 105°=\angle y+30°$
$\angle y=105°-30° \Rightarrow \angle y=75°$

63. (a) $\angle a=\angle b-40°$
(b) $\angle a=\angle b-40° \Rightarrow 70°=\angle b-40°$
$\angle b=70°+40° \Rightarrow \angle y=110°$

65. $2^2+(5-9)=x+3^3$
$4+(-4)=x+27$
$0-27=x+27-27=x+0$
$x=-27$

67. $6x+3-2x=3x-6$
$4x+3-3x-3=3x-6-3x-3$
$x+0=-9 \Rightarrow x=-9$

Cumulative Review Problems

69. Seven times x: $7x$

71. Eight times what number equals forty?
$8n=40$

73. $(15\times 2)+(8\times 1.5)+(6\times -1)+(5\times 2)$
$+(5\times 2)+(2\times -1)=54$ points

3.2 Exercises

1. To solve the equation $-22x=66$, we <u>divide</u> both sides of the equation by <u>-22</u>.

3. $\frac{5x}{?}=x \Rightarrow ?=5$

5. $\frac{-2x}{?}=x \Rightarrow ?=-2$

7. $\frac{6\cdot x}{?}=x \Rightarrow ?=6$

9. $\frac{-1\cdot x}{?}=x \Rightarrow ?=-1$

11. $\frac{3x}{3}=\frac{39}{3} \Rightarrow x=13$

check: $3(13)\overset{?}{=}39,\ 39=39$

13. $\frac{11x}{11}=\frac{55}{11} \Rightarrow x=5$

check: $11(5)\overset{?}{=}55,\ 55=55$

15. $\frac{7y}{7}=\frac{-21}{7}\Rightarrow y=-3$

check: $7(-3)\stackrel{?}{=}-21,\ -21=-21$

17. $\frac{5m}{5}=\frac{-35}{5}\Rightarrow m=-7$

check: $5(-7)\stackrel{?}{=}-35,\ -35=-35$

19. $\frac{-3y}{-3}=\frac{15}{-3}\Rightarrow y=-5$

check: $-3(-5)\stackrel{?}{=}15,\ 15=15$

21. $\frac{-7a}{-7}=\frac{49}{-7}\Rightarrow a=-7$

check: $-7(-7)\stackrel{?}{=}49,\ 49=49$

23. $\frac{42}{6}=\frac{6x}{6}\Rightarrow x=7$

check: $42\stackrel{?}{=}6(7),\ 42=42$

25. $\frac{18}{9}=\frac{9x}{9}\Rightarrow x=2$

check: $18\stackrel{?}{=}9(2),\ 18=18$

27. $\frac{8x}{8}=\frac{104}{8}\Rightarrow x=13$

check: $8(13)\stackrel{?}{=}104,\ 104=104$

29. $\frac{-15y}{-15}=\frac{165}{-15}\Rightarrow y=-11$

check: $-15(-11)\stackrel{?}{=}165,\ 165=165$

31. $\frac{-19x}{-19}=\frac{-76}{-19}\Rightarrow x=4$

check: $-19(4)\stackrel{?}{=}-76,\ -76=-76$

33. $\frac{55}{5}=\frac{5a}{5}\Rightarrow a=11$

check: $55\stackrel{?}{=}5(11),\ 55=55$

35. $-26-18=11a\Rightarrow\frac{11a}{11}=\frac{-44}{11}\Rightarrow a=-4$

check: $-26-18\stackrel{?}{=}11(-4),\ -44=-44$

37. $2x=3-11\Rightarrow\frac{2x}{2}=\frac{-8}{2}\Rightarrow x=-4$

check: $2(-4)\stackrel{?}{=}3-11,\ -8=-8$

39. $-5-5=10y\Rightarrow\frac{-10}{10}=\frac{10y}{10}\Rightarrow y=-1$

check: $-5-5\stackrel{?}{=}10(-1),\ -10=-10$

41. $(3x)\cdot 2=\frac{36}{3}\Rightarrow\frac{6x}{6}=\frac{12}{6}\Rightarrow x=2$

check: $(3\cdot 2)\cdot 2\stackrel{?}{=}\frac{36}{3},\ 12=12$

43. $8x-2x=24\Rightarrow 6x=24\Rightarrow x=\frac{24}{6}=4$

45. $11x-3x=56\Rightarrow 8x=56\Rightarrow x=\frac{56}{8}=7$

47. $65=14x-9x\Rightarrow 5x=65\Rightarrow x=\frac{65}{5}=13$

49. $2(3x)=54\Rightarrow 6x=54\Rightarrow x=\frac{54}{6}=9$

51. $5(4x)=40\Rightarrow 20x=40\Rightarrow x=\frac{40}{20}=2$

53. $5(x\cdot 1)=15\Rightarrow 5x=15\Rightarrow x=\frac{15}{5}=3$

55. $4(x \cdot 2) = 96 \Rightarrow 8x = 96 \Rightarrow x = \dfrac{96}{8} = 12$

57. $L = 4W$

59. $R = 2S$

61. (a) $R = 4W$

(b) $64 = 4W \Rightarrow W = \dfrac{64}{4} = 16$ white roses

63. (a) $L = 2S$

(b) $84 = 2S \Rightarrow S = \dfrac{84}{2} = 42$ in.

65. $35x - 75x = -800 \Rightarrow -40x = -800$

$x = \dfrac{-800}{-40} = 20$ shares of stock

67. $8x + 6x + 9x = 1035 \Rightarrow 23x = 1035$

$x = \dfrac{1035}{23} = \$45$ per hr

69. $x + 5x = 180° \Rightarrow 6x = 180°$

$x = \dfrac{180°}{6} = 30°$

71. $x + 4x = 180° \Rightarrow 5x = 180°$

$x = \dfrac{180°}{5} = 36°$

73. $2550x = 15{,}300 \Rightarrow x = \dfrac{15{,}300}{2550} = 6$

75. $46{,}621x = 466{,}210 \Rightarrow x = \dfrac{466{,}210}{46{,}621} = 10$

77. (a) $13x = 26 \Rightarrow x = \dfrac{26}{13} = 2$

(b) $-13x = 26 \Rightarrow x = \dfrac{26}{-13} = -2$

(c) $x + 13 = 26 \Rightarrow x = 26 - 13 = 13$

(d) $x - 13 = 26 \Rightarrow x = 26 + 13 = 39$

79. (a) $13x = -26 \Rightarrow x = \dfrac{-26}{13} = -2$

(b) $-13x = -26 \Rightarrow x = \dfrac{-26}{-13} = 2$

(c) $x + 13 = -26 \Rightarrow x = -26 - 13 = -39$

(d) $x - 13 = -26 \Rightarrow x = -26 + 13 = -13$

Cumulative Review Exercises

81. $P = 2L + 2W = 2(6) + 2(4)$

$P = 20$ ft

83. $2xy\big|_{x=7,y=9} = 2(7)(9) = 126$

85. $1342 - 562 - 487 = 293$ mi

3.3 Exercises

1. Volume, because we want to find the amount of space inside the pool.

3. $P = 2L + 2W = 2(2) + 2(7)$

$P = 18$ ft

5. $P = 4s = 4(8)$

$P = 32$ ft

7. $P = 4s = 4(54)$

$P = 216$ yd

9. $P = 4s \Rightarrow 36 = 4s \Rightarrow s = \dfrac{36}{4} = 9$ ft

11. $41 = x + 13 + 15 = x + 28$

$x = 41 - 28 = 13$ cm

13. $P = 2L + 2W = 2(4W) + 2W = 10W$

$10W = 30 \Rightarrow W = \frac{30}{10} = 3 \text{ ft}$

15. $P = 2L + 2W = 2(10W) + 2W = 22W$

$22W = 44 \Rightarrow W = \frac{44}{22} = 2 \text{ ft}$

$L = 10W = 10(2) = 20 \text{ ft}$

17. 50 squares

19. 40 squares

21. $A = LW = 22(18) = 396 \text{ ft}^2$

23. $A = s^2 = 8^2 = 64 \text{ in.}^2$

25. $A = bh = 12(9) = 108 \text{ ft}^2$

27. $A = bh = 87(57) = 4959 \text{ m}^2$

29. $A = 60 = 10x \Rightarrow x = \frac{60}{10} = 6 \text{ ft}$

31. $A = 88 = 8h \Rightarrow h = \frac{88}{8} = 11 \text{ m}$

33. $A = 13(19) + 4(3) = 259 \text{ in.}^2$

35. $A = 22(9) + 8(22 - 16) + 7(14) = 344 \text{ m}^2$

37. (a) $4(5)(2) = 40$ cubes

(b) 40 in.^3

39. $V = LWH = 6(5)(9) = 270 \text{ in.}^3$

41. $V = LWH = 27(10)(16) = 4320 \text{ yd}^3$

43. $V = 300 = 10W(6) \Rightarrow W = \frac{300}{60} = 5 \text{ cm}$

45. $V = 200 = L(5)(8) \Rightarrow L = \frac{200}{40} = 5 \text{ m}$

47. $V = LWH = 2(4)(3) = 24 \text{ ft}^3$

49. (a) $2 \text{ ft}\left(\frac{12 \text{ in.}}{\text{ft}}\right) = 24 \text{ in.}$

Dimensions are 24 in. by 24 in.

(b) $A = s^2 = 24^2 = 576 \text{ in.}^2$

51. (a) $4 \text{ yd}\left(\frac{3 \text{ ft}}{\text{yd}}\right) = 12 \text{ ft}, 3 \text{ yd}\left(\frac{3 \text{ ft}}{\text{yd}}\right) = 9 \text{ ft}$

Dimensions are 12 ft by 9 ft.

(b) $A = LW = 12(9) = 108 \text{ ft}^2$

$A = 108 \text{ ft}^2$

53. (a) $12 \text{ ft}\left(\frac{\text{yd}}{3\text{ft}}\right) = 4 \text{ yd}, 9 \text{ ft}\left(\frac{\text{yd}}{3\text{ft}}\right) = 3 \text{ yd}$

$A = LW = 4(3)$

$A = 12 \text{ yd}^2$

(b) $12 \text{ yd}^2\left(\frac{\$8}{\text{yd}^2}\right) = \$96$

55. $(5 \text{ yd})(4 \text{ yd})\left(\frac{9 \text{ ft}^2}{\text{yd}^2}\right)\left(\frac{\text{bag}}{100 \text{ ft}^2}\right)$

$= 1.8 \text{ bags} \rightarrow 2 \text{ bags}\left(\frac{\$3}{\text{bag}}\right) = \$6$

57. $A = 5(7) - 3(2)$

$A = 29 \text{ in.}^2$

59. $A = 9(4) - 2^2 = 36 - 4$

$A = 32 \text{ m}^2$

61. $16(8)-2(4)+22(8)-6(7)+16(8)$
$-4(3)+22(8)-7(3)+22(16)=877\text{ ft}^2$
$877=2(400)+77 \to 2$ gal and 1 qt

Cumulative Review Problems

63. $3426+2,510,777=2,514,203$
$2,514,203=2,514,000$ to nearest thousand

65. $(2)(3x)(5)=6x(5)=5(6x)=30x$

67. $-8\ °\text{F}+21\ °\text{F}=13\ °\text{F}$

3.4 Exercises

1. No, because we do not multiply the bases.

3. Since multiplication is commutative the order in which we multiply does not affect the answer.

5. (a) $(z\cdot z\cdot z)(z\cdot z)=z^3\cdot z^2=z^{3+2}=z^5$
(b) $z^{3+2}=z^5$

7. (a) $(x\cdot x)(x\cdot x\cdot x\cdot x)=x^2\cdot x^4=x^{2+4}=x^6$
(b) $x^{2+4}=x^6$

9. (a) $(z\cdot z)z=z^2\cdot z=z^{2+1}=z^3$
(b) $z^2\cdot z=z^{2+1}=z^3$

11. $x^7\cdot x^2=x^{7+2}=x^9$

13. $x^4\cdot x=x^{4+1}=x^5$

15. $3^2\cdot 3^3=3^{2+3}=3^5$

17. $4\cdot 4^5=4^{1+5}=4^6$

19. $8^2\cdot 7^5=8^2\cdot 7^5$

21. $x^5\cdot y^3=x^5\cdot y^3$

23. $x^5\cdot x^2\cdot x^7=x^{5+2+7}=x^{14}$

25. $y^2\cdot y^4\cdot y^5=y^{2+4+5}=y^{11}$

27. $3^3\cdot 3^2\cdot 3^5=3^{3+2+5}=3^{10}$

29. $2^5\cdot 3^2\cdot 4^7=2^5\cdot 3^2\cdot 4^7$

31. $(4y^5)(6y^7)=24y^{5+7}=24y^{12}$

33. $(6a^6)(9a^8)=54a^{6+8}=54a^{14}$

35. $(-3x)(8x)=-24x^2$

37. $(-4y)(3y)=-12y^2$

39. $(9a)(8a^3)(2a^6)=144a^{1+3+6}=144a^{10}$

41. $(x)(4x^6)(7x^5)=28x^{1+6+5}=28x^{12}$

43. $(5x)(3y)(-2x)=-30x^2y$

45. (a) $5z^3+4$: binomial
(b) 9: monomial
(c) $2x^7-3x^4-3$: trinomial

47. $2x(x^3+6)=2x^4+12x$

49. $6x^2(3x^3-1)=18x^5-6x^2$

51. $5y^3(y^4-2y)=5y^7-10y^4$

53. $-2x^3(x^2+4x)=-2x^5-8x^4$

55. $(2y-5)(-6y^2)=-12y^3+30y^2$

57. $(2x^4 - 4x)(7x^2) = 14x^6 - 28x^3$

59. $A = (2x^4 - 5)(x^3) = 2x^7 - 5x^3$

61. $A = (7x^3 - 4)(x^2) = 7x^5 - 4x^2$

63. $V = 2x^4(7x^3)(x^2) = 14x^9$

65. $P = 2(4x^3) + 2(3x^2) = 8x^3 + 6x^2$

67. $(43x^2)(85x^4)(23x^5) = 84,065x^{11}$

69. $(89x)(-198x^6)(-312x^4) = 5,498,064x^{11}$

71. (a) $4cd + 9cd = (4+9)cd = 13cd$
(b) $(9cd)(4cd) = 36c^2d^2$

73. (a) $9ab^5 - 7ab^5 = (9-7)ab^5 = 2ab^5$
(b) $(9ab^5)(7ab^5) = 63ab^{10}$

75. $W = 4 + L^2$

Cumulative Review Problems

77. $700 - 18 = 682$

79. $20,566 \div 312 = 65$ R 286

81. $-1 + 2 + 0 = 1$ over par

Putting Your Skills to Work, Problems for Individual Investigation

1. $(14 \text{ ft})(11 \text{ ft})\left(\frac{\$9}{\text{ft}^2}\right) = \$1386$

2. $(6 \text{ ft})(7 \text{ ft})\left(\frac{\$6}{\text{ft}^2}\right) = \$252$

Putting Your Skills to Work, Problems for Group Investigation and Cooperative Study

1. (a) $18(10) + 4(4) = 196 \text{ ft}^2$
(b) $14(11) + 6(7) + 4(3) = 208 \text{ ft}^2$
(c) $13(9) + [15(11) + 20(13)]$
$+10(11) = 652 \text{ ft}^2$
(d) $652 \text{ ft}^2\left(\frac{\$8}{\text{ft}^2}\right) = \$5216$

2. (a) $196 \text{ ft}^2\left(\frac{\text{yd}^2}{9 \text{ ft}^2}\right) = 21.\overline{7} \text{ yd}^2 = 20 \text{ yd}^2$ to the nearest ten yd^2.
(b) $20 \text{ yd}^2\left(\frac{\$22}{\text{yd}^2}\right) = \$440$
(c) $208 \text{ ft}^2\left(\frac{\$3}{\text{ft}^2}\right) = \$624$

3. $\$5216 + \$440 + \$624 = \$6280$

Chapter 3 Review Problems

1. $x - 9 = 12 \Rightarrow x = 12 + 9 = 21$
check: $21 - 9 \stackrel{?}{=} 12,\ 12 = 12$

2. $x + 3 = -7 \Rightarrow x = -7 - 3 = -10$
check: $-10 + 3 \stackrel{?}{=} -7,\ -7 = -7$

3. $-7 = y + 2 \Rightarrow y = -7 - 2 = -9$
check: $-7 \stackrel{?}{=} -9 + 2,\ -7 = -7$

4. $6 - 13 + y = -4 + 1 \Rightarrow -7 + y = -3$
$y = -3 + 7 = 4$
check: $6 - 13 + 4 \stackrel{?}{=} -4 + 1,\ -3 = -3$

5. $4(2-6)=x-2 \Rightarrow 4(-4)=x-2$
$-16=x-2 \Rightarrow x=-16+2=-14$

check: $4(2-6)\overset{?}{=}-14-2,\ -16=-16$

6. $4x-3x-6=6 \Rightarrow x=6+6=12$

check: $4(12)-3(12)-6\overset{?}{=}6,\ 6=6$

7. $\angle a+\angle b=180° \Rightarrow \angle a+81°=180°$
$\angle a=180°-81°=99°$

8. $\angle x+\angle y=180° \Rightarrow 21°+\angle y=180°$
$\angle=180°-21°=159°$

9. (a) $\angle x=\angle y+30°$
(b) $101°=\angle y+30° \Rightarrow \angle y=101°-30°$
$\angle y=71°$

10. $\frac{4y}{4}=\frac{48}{4} \Rightarrow y=12$

check: $4(12)\overset{?}{=}48,\ 48=48$

11. $\frac{-8y}{-8}=\frac{56}{-8} \Rightarrow y=-7$

check: $-8(-7)\overset{?}{=}56,\ 56=56$

12. $\frac{6a}{6}=\frac{-42}{6} \Rightarrow a=-7$

check: $6(-7)\overset{?}{=}-42,\ -42=-42$

13. $5(3x)=45 \Rightarrow \frac{15x}{15}=\frac{45}{15} \Rightarrow x=3$

check: $5(3\cdot 3)\overset{?}{=}45,\ 45=45$

14. $3(y\cdot 4)=24 \Rightarrow \frac{12y}{12}=\frac{24}{12} \Rightarrow y=2$

check: $3(2\cdot 4)\overset{?}{=}24,\ 24=24$

15. $6(x\cdot 3)=\frac{72}{2} \Rightarrow \frac{18x}{18}=\frac{36}{18} \Rightarrow x=2$

check: $6(2\cdot 3)\overset{?}{=}\frac{72}{2},\ 36=36$

16. $\frac{-48}{2}=2(3\cdot x) \Rightarrow \frac{-24}{6}=\frac{6x}{6} \Rightarrow x=-4$

check: $\frac{-48}{2}\overset{?}{=}2(3(-4)),\ -24=-24$

17. $-4(5x)=60 \Rightarrow \frac{-20x}{-20}=\frac{60}{-20} \Rightarrow x=-3$

check: $-4(5(-3))\overset{?}{=}60,\ 60=60$

18. $7(2y)=28 \Rightarrow \frac{14y}{14}=\frac{28}{14} \Rightarrow y=2$

check: $7(2\cdot 2)\overset{?}{=}28,\ 28=28$

19. $6x-4x=20 \Rightarrow \frac{2x}{2}=\frac{20}{2} \Rightarrow x=10$

check: $6\cdot 10-4\cdot 10\overset{?}{=}20,\ 20=20$

20. $5x+2x=42 \Rightarrow \frac{7x}{7}=\frac{42}{7} \Rightarrow x=6$

check: $5\cdot 6+2\cdot 6\overset{?}{=}42,\ 42=42$

21. (a) $L=3W$
(b) $\frac{30}{3}=\frac{3W}{3} \Rightarrow W=10$ ft

22. (a) $W=2B$
(b) $W=2B=2(100)=200$

23. $(7+7+9)x=1150 \Rightarrow 23x=1150$
$x=\frac{1150}{23}=\$50$ per hour

24. $x+2x=12 \Rightarrow 3x=12 \Rightarrow x=\frac{12}{3}=4$ mi

25. $P=28=x+8+12 \Rightarrow x+20=28$
$x=28-20 \Rightarrow x=8$ in.

26. $P=2L+2W \Rightarrow 32=2(3W)+2W$

$8W=32 \Rightarrow W=\frac{32}{8} \Rightarrow W=4$ ft

27. $A=s^2=6^2=36\text{ ft}^2$

28. $A=LW=72(52)=3744\text{ in.}^2$

29. $A=bh=9(11)=99\text{ in.}^2$

30. $A=24(11)+10(7)+9(16)$
$A=478\text{ m}^2$

31. $A=104=13x \Rightarrow x=\frac{104}{13}=8$ cm

32. $A=96=x\cdot 8 \Rightarrow x=\frac{96}{8}=12$ in.

33. $V=108=9(6)\cdot H=54H \Rightarrow H=\frac{108}{54}$

$H=2$ ft

34. $V=198=11\cdot W\cdot 6=66W \Rightarrow W=\frac{198}{66}$

$W=3$ in.

35. $V=LWH=25(15)(2)$
$V=750\text{ ft}^3$

36. $V=s^3=5^3$
$V=125\text{ in.}^3$

37. $(16\text{ yd})(12\text{ yd})\left(\frac{9\text{ ft}^2}{\text{yd}^2}\right)=1728\text{ ft}^2$

$1728\text{ ft}^2\left(\frac{\text{bag}}{1000\text{ ft}^2}\right)=1.728\approx 2\text{ bags}$

$\left(\frac{\$4}{\text{bag}}\right)(2\text{ bags})=\$8$

38. (a) $(18\text{ ft})(12\text{ ft})\left(\frac{\text{yd}^2}{9\text{ ft}^2}\right)=24\text{ yd}^2$

(b) $24\text{ yd}^2\left(\frac{\$15}{\text{yd}^2}\right)=\$360$

39. $a^4\cdot a^5=a^{4+5}=a^9$

40. $3^4\cdot 3^6=3^{4+6}=3^{10}$

41. $2^4\cdot 4^3=2^4\cdot 4^3$
or $2^4\cdot(4)(4)(4)=2^4(2^2)(2^2)(2^2)=2^{10}$

42. $x^4\cdot x=x^{4+1}=x^5$

43. $(3y^4)(5y^4)=15y^{4+4}=15y^8$

44. $(4x^2)(3x^6)=12x^{2+6}=12x^8$

45. $(3a)(a^3)(7a^6)=21a^{1+3+6}=21a^{10}$

46. $(4z)(y^8)(3z^3)(2y^3)=24zy^{8+3}=24zy^{11}$

47. $(4x^5)(-3x^2)=-12x^{5+2}=-12x^7$

48. $(-7z^7)(5z^3)=-35z^{7+3}=-35z^{10}$

49. $(-3a^4)(-4a^{10})=12a^{4+10}=12a^{14}$

50. $(3y^4)(-6y^{11})=-18y^{4+11}=-18y^{15}$

51. $3x^2-1$: binomial

52. $4x^3$: monomial

53. $2xy^2+3x+1$: trinomial

54. $x(x^2+2)=x\cdot x^2+x\cdot 2=x^3+2x$

55. $5x(x^3-4)=5x\cdot x^3-5x(4)=5x^4-20x$

56. $A=LW=(3x^2+6)\cdot x^3=3x^5+6x^3$

57. $V=LWH=2x^4\cdot 3x^2\cdot 2x=12x^7$

Chapter 3 Test

1. $x-2=-8\Rightarrow x=-8+2=-6$

check: $-6-2\stackrel{?}{=}-8,\ -8=-8$

2. $y+3=72\Rightarrow y=72-3=69$

check: $69+3\stackrel{?}{=}72,\ 72=72$

3. $9-15=a-3\Rightarrow a=-6+3=-3$

check: $9-15\stackrel{?}{=}-3-3,\ -6=-6$

4. $3x-2x-7=1\Rightarrow x=1+7=8$

check: $3(8)-2(8)-7\stackrel{?}{=}1,\ 1=1$

5. $\frac{5x}{5}=\frac{35}{5}\Rightarrow x=7$

check: $5(7)\stackrel{?}{=}35,\ 35=35$

6. $-3y=42\Rightarrow y=\frac{42}{-3}=-14$

check: $-3(-14)\stackrel{?}{=}42,\ 42=42$

7. $2(4x)=-72\Rightarrow 8x=-72$

$x=\frac{-72}{8}=-9$

check: $2(4(-9))\stackrel{?}{=}-72,\ -72=-72$

8. $\frac{-16}{2}=2+y\Rightarrow -8=2+y\Rightarrow y=-8-2$

$y=-10$

check: $\frac{-16}{2}\stackrel{?}{=}2+(-10),\ -8=-8$

9. $\angle x+\angle y=180°\Rightarrow 75°+\angle y=180°$

$\angle y=180°-75°=105°$

10. (a) $\angle x=\angle y+40°$

(b)

$95°=\angle y+40°\Rightarrow \angle y=95°-40°=55°$

11. (a) $F=3M$

(b)

$21=3M\Rightarrow M=\frac{21}{3}=7$ male students

12. $P=48=2L+2W=2(3W)+2W$

$6W+2W=8W=48\Rightarrow W=\frac{48}{8}=6$ yd

13. $A=LW=5(3)=15\text{ ft}^2$

14. $A=bh=8(9)=72\text{ cm}^2$

15. $V=288=LWH=12W(8)=96W$

$W=\frac{288}{96}=3$ in.

16. $A=LW=7(5)=35\text{ ft}^2$

17. $V=LWH=30(40)(10)=12{,}000\text{ ft}^3$

18. $y^3 \cdot y^2 = y^{3+2} = y^5$

19. $z \cdot z^3 = z^{1+3} = z^4$

20. $a^3b^2 = a^3b^2$

21. $(5x)(x^3)(x^4) = 5x^{1+3+4} = 5x^8$

22. $(-8x^2)(-9x^4) = 72x^{2+4} = 72x^6$

23. $5y(y^4 + 8) = 5y \cdot y^4 + 5y \cdot 8 = 5y^5 + 40y$

24. $6x^3(3x^2 - 5x) = 6x^3 \cdot 3x^2 - 6x^3 \cdot 5x$
$= 18x^5 - 30x^4$

25. $P = 2L + 2W = 2(4x^3) + 2(2x^2)$
$P = 8x^3 + 4x^2$

Cumulative Test for Chapters 1-3

1. $3982 = 4000$ to nearest thousand

2. $1650 - 1475 = 175$ years

3. $x + 9 = 16$

4. $5 + 9y + 3 + y = 5 + 3 + 9y + y = 8 + 10y$

5. $(219)(67) = 14,673$

6. $8446 \div 41 = 206$

7. (a) $167,350 - 86,000 = \$81,350$
(b) $81,350 \div 2 = \$40,675$

8. $-1 > -100$

9. $|-3| = 3,\ |-2| = 2,\ 3 > 2 \Rightarrow |-3| > |-2|$

10. $(5) + (-5) = 0$

11. $-10 - 81 = -10 + (-81) = -91$

12. $(16)(-4) = -64$

13. $(-36) \div (-9) = 4$

14. $(-1)^5 = (-1)(-1)(-1)(-1)(-1) = -1$

15. $25 + x = 85 \Rightarrow x = 85 - 25 = 60$

16. $9y = 63 \Rightarrow y = \frac{63}{9} = 7$

17. $-36 + y = -2 \Rightarrow y = -2 + 36 = 34$

18. $2x - x + 5 = -6 \Rightarrow x = -6 - 5 = -11$

19. $\frac{120}{10} = -3y \Rightarrow -3y = 12 \Rightarrow y = \frac{12}{-3} = -4$

20. $2 + 6 = -3 + 4 + x \Rightarrow 8 = 1 + x$
$x = 8 - 1 = 7$

21. $P = 2L + 2W = 2(3) + 2(2) = 10$ in.

22. $V = LWH,\ 30 = L(2)(3) \Rightarrow 6L = 30$
$L = \frac{30}{6} = 5$ m

23. $\angle a + \angle b = 180° \Rightarrow \angle a + 105° = 180°$
$\angle a = 180° - 105° = 75°$

24. $A = s^2 = 12^2 = 144$ ft^2

25. $(-3x^2)(4x^6) = -12x^{2+6} = -12x^8$

26. $(5y^2)(3y^5) = 15y^{2+5} = 15y^7$

27. $x^3(2x^2+5) = x^3 \cdot 2x^2 + x^3 \cdot 5$
$= 2x^5 + 5x^3$

28. $A = LW = (3x^3 - 1)(2x^2)$
$A = 3x^3 \cdot 2x^2 - 1 \cdot 2x^2$
$A = 6x^5 - 2x^2$

Chapter 4

Pretest Chapter 4

1. 216 is even and thus divisible by 2. 216 is divisible by 3 since the sum of its digits, $2+1+6=9$, is divisible by 3. 216 is not divisible by 5 because the last digit is not 0 or 5.

2. (a) 0 is neither prime nor composite.
(b) 17 is prime
(c) Since $142 = 2\cdot 71$, 142 is composite.

3. $120 = 2\cdot 2\cdot 2\cdot 3\cdot 5 = 2^3\cdot 3\cdot 5$

4. $\dfrac{7}{10}$

5. (a) $5\dfrac{2}{3}$: mixed number

(b) $\dfrac{6}{7}$: proper fraction

(c) $\dfrac{9}{3}$: improper fraction

6. $\dfrac{12}{5} = 2\dfrac{2}{5}$

7. $2\dfrac{3}{4} = \dfrac{4\cdot 2+3}{4} = \dfrac{8+3}{4} = \dfrac{11}{4}$

8. $\dfrac{3}{5}\cdot\dfrac{2}{2} = \dfrac{6}{10}$

9. $\dfrac{5}{12}\cdot\dfrac{5}{5} = \dfrac{25}{60}$

10. $\dfrac{-24}{44} = \dfrac{-\cancel{2}\cdot\cancel{2}\cdot 2\cdot 3}{\cancel{2}\cdot\cancel{2}\cdot 11} = \dfrac{-6}{11}$

11. $(y^2)^4 = y^{2\cdot 4} = y^8$

12. $\dfrac{16x^7}{24x^3} = \dfrac{\cancel{2}\cdot\cancel{2}\cdot\cancel{2}\cdot 2x^{7-3}}{\cancel{2}\cdot\cancel{2}\cdot\cancel{2}\cdot 3} = \dfrac{2x^4}{3}$

13. $\left(\dfrac{x}{5}\right)^2 = \left(\dfrac{x}{5}\right)\left(\dfrac{x}{5}\right) = \dfrac{x^2}{25}$

14. $\dfrac{\text{males}}{\text{females}} = \dfrac{16}{28} = \dfrac{\cancel{2}\cdot\cancel{2}\cdot 2\cdot 2}{\cancel{2}\cdot\cancel{2}\cdot 7} = \dfrac{4}{7}$

15. $\dfrac{25 \text{ lb}}{5 \text{ \$}} = \dfrac{\cancel{5}\cdot 5 \text{ lb}}{\cancel{5} \text{ \$}} = 5$ lb per dollar

16. $\dfrac{364 \text{ mi}}{14 \text{ gal}} = \dfrac{\cancel{2}\cdot 2\cdot\cancel{7}\cdot 13 \text{ mi}}{\cancel{2}\cdot\cancel{7} \text{ gal}} = 26 \text{ mi/gal}$

17. $4\cdot 36 = 144,\ 9\cdot 18 = 162,$
not a proportion

18. $\dfrac{x}{5} = \dfrac{8}{20} \Rightarrow 20x = 40 \Rightarrow x = 2$

19. $\dfrac{1 \text{ lb}}{250 \text{ ft}^2}\cdot\dfrac{7}{7} = \dfrac{7 \text{ lb}}{1750 \text{ ft}^2}$

4.1 Exercises

1. With this method the divisors and quotient are the prime factors. Therefore, the divisor (number you divide by) must be prime for your factors to be prime.

3. 155 is not divisible by 2 because it is not even.

5. 232 is not divisible by three because the sum of the digits, $2+3+2=7$, is not divisible by 3.

7. 324 is even and therefore divisible by 2. The sum of the digits, $3+2+4=9$, is divisible by 3 and therefore 324 is divisible by 3. The last digit is not a 0 or 5, therefore 324 is not divisible by 5.

9. 805 is not even and therefore not divisible by 2. The sum of the digits, $8+0+5=13$, is not divisible by 3 and therefore 805 is not divisible by 3. The last digit is 5, therefore 805 is divisible by 5.

11. 330 is even and therefore divisible by 2. The sum of the digits, $3+3+0=6$, is divisible by 3 and therefore 330 is divisible by 3. The last digit is 0, therefore 330 is divisible by 5.

13. 22,971 is not even and therefore is not divisible by 2. The sum of the digits, $2+2+9+7+1=21$, is divisible by 3 and therefore 22,971 is divisible by 3. The last digit is not a 0 or 5, therefore 22,971 is not divisible by 5.

15. 991,500 is even and therefore divisible by 2. The sum of the digits, $9+9+1+5=24$, is divisible by 3 and therefore 991,500 is divisible by 3. The last digit is 0, therefore 991,500 is divisible by 5.

17. 0,1: neither. 23: prime.
9,40,8,13,33: composite.

19. (a) $56=8\cdot 7$
(b) $56=2\cdot 2\cdot 2\cdot 7=2^3\cdot 7$

21. $9=3\cdot 3=3^2$

23. $12=2\cdot 2\cdot 3=2^2\cdot 3$

25. $24=2\cdot 2\cdot 2\cdot 3=2^3\cdot 3$

27. $36=2\cdot 2\cdot 3\cdot 3=2^2\cdot 3^2$

29. $64=2\cdot 2\cdot 2\cdot 2\cdot 2\cdot 2=2^6$

31. $80=2\cdot 2\cdot 2\cdot 2\cdot 5=2^4\cdot 5$

33. $75=3\cdot 5\cdot 5=3\cdot 5^2$

35. $45=3\cdot 3\cdot 5=3^2\cdot 5$

37. $99=3\cdot 3\cdot 11=3^2\cdot 11$

39. $105=3\cdot 5\cdot 7$

41. $110=2\cdot 5\cdot 11$

43. $136=2\cdot 2\cdot 2\cdot 17=2^3\cdot 17$

45. $220=2\cdot 2\cdot 5\cdot 11=2^2\cdot 5\cdot 11$

47. $1309=7\cdot 11\cdot 17$

49. $2737=7\cdot 17\cdot 23$

51. The sum of the digits is divisible by 9.
$2{,}148{,}543=238{,}727\cdot 9$
$2+1+4+8+5+4+3=27=3\cdot 9$

53. $27{,}000=3\cdot 9{,}000=5\cdot 5400$. Other answers are possible but the number must end in 5 or 0 and the sum of the digits must be divisible by 3.

Cumulative Review Problems

55. $(2x^2)(5x^3y)=10x^{2+3}y=10x^5y$

57. $5x+3x+2=(5+3)x+2=8x+2$

59. $9(325+220)=9(545)=\$4905$

4.2 Exercises

1. To change a mixed number to an improper fraction, we multiply the denominator times the whole number and then add the numerator to this product. The denominator of the improper fraction is the same as the one in the mixed number.

3. $\frac{3}{5}$

5. $\frac{3}{4}$

7. $\frac{7}{0}$ undefined

9. $\frac{0}{z}=0$

11. $\frac{44}{44}=1$

13. $\frac{y}{y}=1$

15. $\frac{8}{0}$ undefined

17. $\frac{0}{8}=0$

19. $\frac{7}{15}$

21. $\frac{37}{37+57}=\frac{37}{94}$

23. $\frac{26-9}{26}=\frac{17}{26}$

25. $\frac{32-21}{32}=\frac{11}{32}$

27. $\frac{58}{24+22+58+31}=\frac{58}{135}$

29. $\frac{31}{24+22+58+31}=\frac{31}{135}$

31. $\frac{7}{9}$: proper fraction

33. $\frac{9}{9}$: improper fraction

35. $\frac{6}{6}$: improper fraction

37. $7\frac{1}{9}$: mixed number

39. $\frac{13}{7}=1\frac{6}{7}$

41. $\frac{73}{4}=18\frac{1}{4}$

43. $\frac{41}{2}=20\frac{1}{2}$

45. $\frac{32}{5}=6\frac{2}{5}$

47. $\frac{47}{5}=9\frac{2}{5}$

49. $\frac{44}{44}=1$

51. $8\frac{3}{7}=\frac{8(7)+3}{7}=\frac{56+3}{7}=\frac{59}{7}$

53. $24\frac{1}{4}=\frac{4(24)+1}{4}=\frac{96+1}{4}=\frac{97}{4}$

55. $15\frac{2}{3}=\frac{3(15)+2}{3}=\frac{45+2}{3}=\frac{47}{3}$

57. $33\frac{1}{3}=\frac{3(33)+1}{3}=\frac{99+1}{3}=\frac{100}{3}$

59. $8\frac{9}{10}=\frac{10(8)+9}{10}=\frac{80+9}{10}=\frac{89}{10}$

61. $8\frac{21}{30}=\frac{30(8)+21}{30}=\frac{240+21}{30}=\frac{261}{30}$

Cumulative Review Problems

63. $(-5)(-8)=40$

65. $63=3\cdot 3\cdot 7=3^2\cdot 7$

67. $-71\text{ °C}-(-281\text{ °C})=271\text{ °C}$

4.3 Exercises

1. To build an equivalent fraction we <u>multiply</u> the numerator and denominator by the <u>same</u> number.

3. (a) $\frac{7}{9}\cdot\frac{4}{4}=\frac{28}{36}$

(b) $\frac{7}{9}\cdot\frac{5x}{5x}=\frac{35x}{45x}$

5. (a) $\frac{4}{11}\cdot\frac{4}{4}=\frac{16}{44}$

(b) $\frac{4}{11}\cdot\frac{5x}{5x}=\frac{20x}{55x}$

7. $\frac{3}{7}\cdot\frac{7}{7}=\frac{21}{49}$

9. $\frac{3}{4}\cdot\frac{5}{5}=\frac{15}{20}$

11. $\frac{9}{13}\cdot\frac{3}{3}=\frac{27}{39}$

13. $\frac{35}{40}\cdot\frac{2}{2}=\frac{70}{80}$

15. $\frac{8}{9}\cdot\frac{y}{y}=\frac{8y}{9y}$

17. $\frac{3}{7}\cdot\frac{4y}{4y}=\frac{12y}{28y}$

19. $\frac{3}{6}\cdot\frac{3a}{3a}=\frac{9a}{18a}$

21. $\frac{5}{7}\cdot\frac{3x}{3x}=\frac{15x}{21x}$

23. $\frac{15}{25}=\frac{3\cdot\cancel{5}}{5\cdot\cancel{5}}=\frac{3}{5}$

25. $\frac{12}{16}=\frac{\cancel{2}\cdot\cancel{2}\cdot 3}{\cancel{2}\cdot\cancel{2}\cdot 2\cdot 2}=\frac{3}{4}$

27. $\frac{30}{36}=\frac{\cancel{2}\cdot\cancel{3}\cdot 5}{\cancel{2}\cdot 2\cdot\cancel{3}\cdot 3}=\frac{5}{6}$

29. $\frac{24}{28}=\frac{\cancel{2}\cdot\cancel{2}\cdot 2\cdot 3}{\cancel{2}\cdot\cancel{2}\cdot 7}=\frac{6}{7}$

31. $\frac{24}{36}=\frac{\cancel{2}\cdot\cancel{2}\cdot 2\cdot\cancel{3}}{\cancel{2}\cdot\cancel{2}\cdot\cancel{3}\cdot 3}=\frac{2}{3}$

33. $\frac{30}{85}=\frac{2\cdot 3\cdot\cancel{5}}{\cancel{5}\cdot 17}=\frac{6}{17}$

35. $\frac{42}{54}=\frac{\cancel{2}\cdot\cancel{3}\cdot 7}{\cancel{2}\cdot\cancel{3}\cdot 3\cdot 3}=\frac{7}{9}$

37. $\frac{36}{72}=\frac{\cancel{2}\cdot\cancel{2}\cdot\cancel{3}\cdot\cancel{3}}{\cancel{2}\cdot\cancel{2}\cdot 2\cdot\cancel{3}\cdot\cancel{3}}=\frac{1}{2}$

39. $\frac{49}{35}=\frac{7\cdot\cancel{7}}{5\cdot\cancel{7}}=\frac{7}{5}=1\frac{2}{5}$

41. $\frac{75}{60}=\frac{\cancel{3}\cdot\cancel{5}\cdot 5}{2\cdot 2\cdot\cancel{3}\cdot\cancel{5}}=\frac{5}{4}=1\frac{1}{4}$

43. $\frac{-24}{36}=\frac{-\cancel{2}\cdot\cancel{2}\cdot 2\cdot\cancel{3}}{\cancel{2}\cdot\cancel{2}\cdot\cancel{3}\cdot 3}=-\frac{2}{3}$

45. $\frac{-42}{48}=\frac{-\cancel{2}\cdot\cancel{3}\cdot 7}{\cancel{2}\cdot 2\cdot 2\cdot 2\cdot\cancel{3}}=-\frac{7}{8}$

47. $\frac{30}{-42}=-\frac{\cancel{2}\cdot\cancel{3}\cdot 5}{\cancel{2}\cdot\cancel{3}\cdot 7}=-\frac{5}{7}$

49. $-\frac{16}{18}=-\frac{\cancel{2}\cdot 2\cdot 2\cdot 2}{\cancel{2}\cdot 3\cdot 3}=-\frac{8}{9}$

51. (a) $\frac{-12}{18}=-\frac{\cancel{2}\cdot 2\cdot\cancel{3}}{\cancel{2}\cdot\cancel{3}\cdot 3}=-\frac{2}{3}$

(b) $\frac{12}{-18}=-\frac{\cancel{2}\cdot 2\cdot\cancel{3}}{\cancel{2}\cdot\cancel{3}\cdot 3}=-\frac{2}{3}$

(c) $-\frac{12}{18}=-\frac{\cancel{2}\cdot 2\cdot\cancel{3}}{\cancel{2}\cdot\cancel{3}\cdot 3}=-\frac{2}{3}$

53. $\frac{24xy}{42x}=\frac{\cancel{2}\cdot 2\cdot 2\cdot\cancel{3}\cancel{x}y}{\cancel{2}\cdot\cancel{3}\cdot 7\cancel{x}}=\frac{4y}{7}$

55. $\frac{21x}{28xy}=\frac{3\cdot\cancel{7}\cancel{x}}{2\cdot 2\cdot\cancel{7}\cancel{x}y}=\frac{3}{4y}$

57. $\frac{27x^2}{45x}=\frac{\cancel{3}\cdot\cancel{3}\cdot 3\cdot\cancel{x}\cdot x}{\cancel{3}\cdot\cancel{3}\cdot 5\cdot\cancel{x}}=\frac{3x}{5}$

59. $\frac{20y}{24y^2}=\frac{\cancel{2}\cdot\cancel{2}\cdot 5\cancel{y}}{\cancel{2}\cdot\cancel{2}\cdot 2\cdot 3\cdot\cancel{y}\cdot y}=\frac{5}{6y}$

61. $\frac{16x}{18x}=\frac{\cancel{2}\cdot 2\cdot 2\cdot 2\cancel{x}}{\cancel{2}\cdot 3\cdot 3\cancel{x}}=\frac{8}{9}$

63. $\frac{20y}{35y}=\frac{2\cdot 2\cdot\cancel{5}\cancel{y}}{\cancel{5}\cdot 7\cancel{y}}=\frac{4}{7}$

65. $\frac{36n^2}{-42n}=-\frac{\cancel{2}\cdot 2\cdot\cancel{3}\cdot 3\cdot\cancel{n}\cdot n}{\cancel{2}\cdot\cancel{3}\cdot 7\cancel{n}}=-\frac{6n}{7}$

67. $\frac{-35x}{45x^2}=-\frac{\cancel{5}\cdot 7\cancel{x}}{3\cdot 3\cdot\cancel{5}\cancel{x}\cdot x}=-\frac{7}{9x}$

69. $\frac{8}{24}=\frac{\cancel{8}}{\cancel{8}\cdot 3}=\frac{1}{3}$

71. $\frac{36-22}{36}=\frac{14}{36}=\frac{\cancel{2}\cdot 7}{\cancel{2}\cdot 18}=\frac{7}{18}$

73. $\frac{57}{74+40+57+49+58+79}=\frac{57}{357}=\frac{19}{119}$

75. $\frac{74+40}{357}=\frac{114}{357}=\frac{38}{119}$

77. $\frac{40a^2b^2c^4}{88ab}=\frac{5\cdot\cancel{8}\cancel{a}\cdot a\cdot\cancel{b}\cdot bc^4}{\cancel{8}\cdot 11\cancel{ab}}=\frac{5abc^4}{11}$

79. $\frac{256xy^3}{300yz^5}=\frac{\cancel{4}\cdot 64x\cancel{y}\cdot y^2}{\cancel{4}\cdot 75\cancel{y}z^5}=\frac{64xy^2}{75z^5}$

Cumulative Review Problems

81. $9-(6-1)=9-5=4$

83. $84(48)=168W \Rightarrow W=\frac{4032}{168}=24$ in.

4.4 Exercises

1. (a) We add the coefficients and the variables stay the same.
(b) We multiply coefficients and then add exponents of like terms.
(c) We simplify coefficients and then subtract exponents of like bases.

3. $\frac{7^4}{7^3}=7^{4-3}=7^1=7$

5. $\frac{a^8}{a^3}=a^{8-3}=a^5$

7. $\frac{3^5}{3^9}=3^{5-9}=3^{-4}=\frac{1}{3^4}$

9. $\frac{3^2}{3^6}=3^{2-6}=3^{-4}=\frac{1}{3^4}$

11. $\frac{z^4}{y^8}=\frac{z^4}{y^8}$

13. $\frac{9^3}{8^3}=\frac{9^3}{8^3}$

15. $\frac{z^8}{z^8}=1$

17. $\frac{y^3z^4}{y^5z^7}=y^{3-5}z^{4-7}=y^{-2}z^{-3}=\frac{1}{y^2z^3}$

19. $\frac{m^9 3^6}{m^7 3^7}=m^{9-7}\cdot 3^{6-7}=m^2\cdot 3^{-1}=\frac{m^2}{3}$

21. $\frac{a^5 7^4}{a^3 7^7}=a^{5-3}\cdot 7^{4-7}=a^2\cdot 7^{-3}=\frac{a^2}{7^3}$

23. $\frac{b^9 9^9}{b^7 9^{11} 3^0}=b^{9-7}9^{9-11}=b^2 9^{-2}=\frac{b^2}{9^2}$

25. $\frac{25y^3}{35y}=\frac{\cancel{5}\cdot 5\cancel{y}y^2}{\cancel{5}\cdot 7\cancel{y}}=\frac{5y^2}{7}$

27. $\frac{9a^4}{27a^3}=\frac{\cancel{9}\cancel{a^3}a}{\cancel{9}\cdot 3\cancel{a^3}}=\frac{a}{3}$

29. $\frac{56x^9y^0}{64x^3}=\frac{7\cdot\cancel{8}x^{9-3}\cdot 1}{8\cdot\cancel{8}}=\frac{7x^6}{8}$

31. (a) $z^2\cdot z^2\cdot z^2=z^{2+2+2}=z^6$
(b) $(z^2)^3=z^{2\cdot 3}=z^6$

33. (a) $x^2\cdot x^2=x^{2+2}=x^4$
(b) $(x^2)^2=x^{2\cdot 2}=x^4$

35. (a) $b^4\cdot b^4\cdot b^4=b^{4+4+4}=b^{12}$
(b) $(b^4)^3=b^{4\cdot 3}=b^{12}$

37. $(z^6)^4=z^{6\cdot 4}=z^{24}$

39. $(3^3)^2 = 3^{3\cdot 2} = 3^6$

41. $(b^1)^6 = (b)^6 = b^6$

43. $(x^0)^4 = (1)^4 = 1\cdot 1\cdot 1\cdot 1 = 1$

45. $(y^2)^3 = y^{2\cdot 3} = y^6$

47. $(2^4)^5 = 2^{4\cdot 5} = 2^{20}$

49. $(x^2)^0 = x^{2\cdot 0} = x^0 = 1$

51. $(7^3)^9 = 7^{3\cdot 9} = 7^{27}$

53. $(x^2)^2 = x^{2\cdot 2} = x^4$

55. $(y^2)^6 = y^{2\cdot 6} = y^{12}$

57. $\left(\frac{2}{x}\right)^2 = \left(\frac{2}{x}\right)\left(\frac{2}{x}\right) = \frac{2^2}{x^2} = \frac{4}{x^2}$

59. $\left(\frac{y}{x}\right)^9 = \frac{y^9}{x^9}$

61. $\left(\frac{3}{x}\right)^3 = \frac{3^3}{x^3} = \frac{27}{x^3}$

63. $\left(\frac{a}{b}\right)^4 = \frac{a^4}{b^4}$

65. $\left(\frac{x}{6}\right)^2 = \frac{x^2}{6^2} = \frac{x^2}{36}$

67. $\left(\frac{3}{4}\right)^2 = \frac{3^2}{4^2} = \frac{9}{16}$

69. (a) $15x^3 + 5x^3 = (15+5)x^3 = 20x^3$
(b) $(15x^3)(5x^3) = 75x^{3+3} = 75x^6$
(c) $(x^3)^3 = x^{3\cdot 3} = x^9$
(d) $\frac{15x^3}{5x^5} = 3x^{3-5} = 3x^{-2} = \frac{3}{x^2}$

71. (a) $3x^3 + 9x^3 = (3+9)x^3 = 12x^3$
(b) $(3x^3)(9x^4) = 27x^{3+4} = 27x^7$
(c) $(3x^2)^4 = 3^4 x^{2\cdot 4} = 3^4 x^8$
(d) $\frac{3x^3}{9x^4} = \frac{\cancel{3}\cancel{x^3}}{\cancel{3}\cdot 3\cancel{x^3}\cdot x} = \frac{1}{3x}$

73. $\frac{25x^2y^3z^4}{135x^7y} = \frac{\cancel{5}\cdot 5\cancel{x^2}\cancel{y}y^2z^4}{\cancel{5}\cdot 27\cancel{x^2}x^5\cancel{y}} = \frac{5y^2z^4}{27x^5}$

75. $\frac{156a^0b^8}{144b^6c^9} = \frac{\cancel{12}\cdot 13\cdot 1\cdot b^{8-6}}{\cancel{12}\cdot 12c^9} = \frac{13b^2}{12c^9}$

Cumulative Review Problems

77. $3x = 42 \Rightarrow x = \frac{42}{3} = 14$

79. $(18-4) = 7x \Rightarrow 7x = 14 \Rightarrow x = \frac{14}{7} = 2$

81. $18(18) - 40 = 324 - 40 = 284$ squares

4.5 Exercises

1. A ratio compares amounts with the same units and a rate compares amounts with different units.

3. $\frac{15}{65} = \frac{3\cdot\cancel{5}}{\cancel{5}\cdot 13} = \frac{3}{13}$

5. $\frac{35}{10} = \frac{\cancel{5} \cdot 7}{\cancel{5} \cdot 2} = \frac{7}{2}$

7. $\frac{25}{70} = \frac{\cancel{5} \cdot 5}{\cancel{5} \cdot 14} = \frac{5}{14}$

9. $\frac{34 \text{ min}}{12 \text{ min}} = \frac{\cancel{2} \cdot 17}{\cancel{2} \cdot 6} = \frac{17}{6}$

11. $\frac{14 \text{ gal}}{35 \text{ gal}} = \frac{2 \cdot \cancel{7}}{5 \cdot \cancel{7}} = \frac{2}{5}$

13. $\frac{17 \text{ hr}}{41 \text{ hr}} = \frac{17}{41}$

15. $\frac{\$121}{\$423} = \frac{121}{423}$

17. (a) $\frac{15}{35} = \frac{3 \cdot \cancel{5}}{\cancel{5} \cdot 7} = \frac{3}{7}$

(b) $\frac{35}{15} = \frac{\cancel{5} \cdot 7}{\cancel{5} \cdot 3} = \frac{7}{3}$

19. (a) $\frac{\text{wins}}{\text{losses}} = \frac{23}{14}$ (b) $\frac{\text{losses}}{\text{wins}} = \frac{14}{23}$

21. $\frac{1 \text{ gram}}{4 \text{ gram}} = \frac{1}{4}$

23. $\frac{5328 \text{ ft}}{6066 \text{ ft}} = \frac{1\cancel{8} \cdot 296}{1\cancel{8} \cdot 337} = \frac{296}{337}$

25. $\frac{205 \text{ cal}}{7 \text{ gram fat}} = 29\frac{2}{7}$ cal per gram fat

27. $\frac{405 \text{ mi}}{18 \text{ gal}} = 22\frac{1}{2}$ mi/gal

29. $\frac{\$455}{35 \text{ hr}} = \$13$ per hr

31. $\frac{410 \text{ mi}}{7 \text{ hr}} = 58\frac{4}{7}$ mph

33. $\frac{63 \text{ rotations}}{18 \text{ sec}} = 3\frac{1}{2}$ rotations/sec

35. $\frac{\$189}{21 \text{ CD's}} = \$9$ per CD

37. (a) $\frac{620 \text{ clients}}{4 \text{ agents}} = 155$ clients per agent

(b) $\frac{310 \text{ clients}}{5 \text{ clerical staff}} = 62$ clients per staff

(c) $\frac{310 \text{ clients}}{5 \text{ clerical staff}} \cdot \frac{3}{3} = \frac{930 \text{ clients}}{15 \text{ clerical staff}}$

39. (a) $\frac{\$96}{12 \text{ sessions}} = \$8$ per session

$\frac{\$135}{15 \text{ sessions}} = \$9$ per session

(b) The 12-session special is the better deal.

41. (a) $\frac{\$12}{4 \text{ reams}} = \$3$ per ream

$\frac{\$21}{7 \text{ reams}} = \$3$ per ream

(b) Both offer the same unit price.

43. $\frac{1079}{31} = 34\frac{25}{31} \rightarrow 35$ sales per day

45. $\frac{3800 \text{ mi}}{190 \text{ gal}} = 20 \text{ mi/gal}$

Cumulative Review Problems

47. $15n = 180 \Rightarrow n = \frac{180}{15} = 12$

check: $15(12) \stackrel{?}{=} 180,\ 180 = 180$

49. $48 = 3x \Rightarrow x = \frac{48}{3} = 16$

check: $48 \stackrel{?}{=} 3(16),\ 48 = 48$

51. $-4 + 8 - 5 = -1$, a loss of 1 yard

4.6 Exercises

1. When the numerator and denominator of a fraction are equal, the fraction is equal to 1. Thus both fractions are equal to 1.

3. $7(32) = 224 = 8(28)$

5. $9(66) = 594 \neq 462 = 11(42)$

7. $5(12) = 60 \neq 11(7) = 77$

9. $1(15) = 15 \neq 8(1) = 8$

11. $4(105) = 420 = 21(20)$

13. =, both fractions equal 1

15. $3(16) = 48 = 12(4)$, yes

17. $3(12) = 36 \neq 40 = 8(5)$, no

19. $\frac{4}{9} = \frac{28}{63}$

21. $\frac{12}{7} = \frac{48}{28}$

23. $\frac{3}{8} = \frac{18}{48}$

25. $\frac{\frac{1}{3}}{\frac{1}{8}} = \frac{\frac{1}{4}}{\frac{3}{32}}$

27. $\frac{\text{2 printers}}{\text{4 secrtaries}} = \frac{\text{12 printers}}{\text{24 secrtaries}}$

29. $\frac{\text{2 cups}}{\text{50 grams}} = \frac{\text{6 cups}}{\text{150 grams}}$

31. $\frac{\text{6 American \$}}{\text{4 British pounds}} = \frac{\text{264 American \$}}{\text{176 British pounds}}$

33. $\frac{3\frac{1}{2}\text{ rotations}}{\text{2 min}} = \frac{\text{14 rotations}}{\text{8 min}}$

35. $\frac{\text{4 made}}{\text{7 attempts}} = \frac{\text{12 made}}{\text{21 attempts}}$

37. $2(28) = 56 = 7(8)$, yes

39. $13(29) = 377 \neq 494 = 19(26)$, no

41. $2(135) = 270 \neq 715 = 11(65)$, no

43. $180(6) = 1080 = 3(360)$, yes

45. $\frac{400}{5} \stackrel{?}{=} \frac{675}{9}$

$400(9) = 3600 \neq 3375 = 5(675)$, no

47. $\frac{x}{8} = \frac{5}{2} \Rightarrow 2x = 40 \Rightarrow x = 20$

check: $\frac{20}{8} \stackrel{?}{=} \frac{5}{2},\ \frac{5}{2} = \frac{5}{2}$

49. $\frac{x}{12}=\frac{8}{3}\Rightarrow 3x=96\Rightarrow x=32$

check: $\frac{32}{12}\overset{?}{=}\frac{8}{3},\ \frac{8}{3}=\frac{8}{3}$

51. $\frac{12}{x}=\frac{3}{5}\Rightarrow 3x=60\Rightarrow x=20$

check: $\frac{12}{20}\overset{?}{=}\frac{3}{5},\ \frac{3}{5}=\frac{3}{5}$

53. $\frac{18}{x}=\frac{3}{4}\Rightarrow 3x=72\Rightarrow x=24$

check: $\frac{18}{24}\overset{?}{=}\frac{3}{4},\ \frac{3}{4}=\frac{3}{4}$

55. $\frac{9}{75}=\frac{6}{x}\Rightarrow 9x=450\Rightarrow x=50$

check: $\frac{9}{75}\overset{?}{=}\frac{6}{50},\ \frac{3}{25}=\frac{3}{25}$

57. $\frac{7}{5}=\frac{x}{25}\Rightarrow 5x=175\Rightarrow x=35$

check: $\frac{7}{5}\overset{?}{=}\frac{35}{25},\ \frac{7}{5}=\frac{7}{5}$

59. $\frac{15}{x}=\frac{30}{14}\Rightarrow 30x=210\Rightarrow x=7$

check: $\frac{15}{7}\overset{?}{=}\frac{30}{14},\ \frac{15}{7}=\frac{15}{7}$

61. $\frac{15}{6}=\frac{5}{x}\Rightarrow 15x=30\Rightarrow x=2$

check: $\frac{15}{6}\overset{?}{=}\frac{5}{2},\ \frac{5}{2}=\frac{5}{2}$

63. $\frac{35}{28}=\frac{n}{4}\Rightarrow 28n=140\Rightarrow n=5$

check: $\frac{35}{28}\overset{?}{=}\frac{5}{4},\ \frac{5}{4}=\frac{5}{4}$

65. $\frac{80\text{ gal}}{24\text{ acres}}=\frac{20\text{ gal}}{n\text{ acres}}\Rightarrow 80n=480\Rightarrow$

$n=6$

67. $\frac{10\text{ mi}}{16\text{ km}}=\frac{n\text{ mi}}{8\text{ km}}\Rightarrow 16n=80\Rightarrow n=5$

69. $\frac{n\text{ mi}}{15\text{ gal}}=\frac{16\text{ mi}}{3\text{ gal}}\Rightarrow 3n=240\Rightarrow n=80$

71. $244(322)=78{,}568\neq 74{,}404=356(209)$

no

73. $355(1724)=612{,}020=431(1420)$, yes

75. $\frac{1\text{ day}}{16\text{ million}}\cdot\frac{7}{7}=\frac{7\text{ day}}{112\text{ million}}$

Cumulative Review Problems

77. $2x+6=28$

79. $x-8=9$

81. $5(-1)+50(5)+10(-3)=215$ points

4.7 Exercises

1. (a) $\frac{24}{4}=6=\frac{36}{6}$, yes

(b) $\frac{24}{3}=8\neq\frac{36}{5}=7\frac{1}{5}$, no

3. $\frac{2\text{ cups}}{50\text{ g}}=\frac{5\text{ cups}}{x}\Rightarrow 2x=250,\ x=125\text{ g}$

5. $\frac{3\text{ shirts}}{\$25}=\frac{12\text{ shirts}}{c}\Rightarrow 3c=300$

$c=\$100$

7. $\frac{25 \text{ min}}{2 \text{ rows}} = \frac{t}{12 \text{ rows}} \Rightarrow 2t = 300$

$t = 150 \text{ min}$

9. $\frac{210 \text{ ft}}{3 \text{ in.}} = \frac{x}{5 \text{ in.}} \Rightarrow 3x = 1050,\ x = 350 \text{ ft}$

11. $\frac{5 \text{ shares}}{\$160} = \frac{12 \text{ shares}}{c} \Rightarrow 5c = 1920$

$c = \$384$

13. $\frac{5 \text{ cups water}}{2 \text{ cups punch}} = \frac{x}{8 \text{ cups punch}}$

$2x = 40 \Rightarrow x = 20 \text{ cups water}$

15. $\frac{8 \text{ shares}}{5 \text{ held}} = \frac{n}{850 \text{ held}} \Rightarrow 5n = 6800$

$n = 1360 \text{ shares}$

17. $\frac{100 \text{ watt}}{30 \text{ mm}} = \frac{140 \text{ watt}}{t} \Rightarrow 100t = 4200$

$t = 42 \text{ mm}$

19. $\frac{12 \text{ ft}}{18 \text{ ft}} = \frac{\text{W}}{30 \text{ ft}} \Rightarrow 18W = 360,\ W = 20 \text{ ft}$

21. $\frac{\$325}{\$1950} = \frac{\$40}{x} \Rightarrow 325x = 78{,}000$

$x = \$240$

23. $\frac{\$325}{\$1950} = \frac{\$32}{x} \Rightarrow 325x = 62{,}400$

$x = \$192$

25. $\frac{\$325}{\$1950} = \frac{\$215}{x} \Rightarrow 325x = 419{,}250$

$x = \$1290$

27. $\frac{6}{7} = \frac{\$2400}{K} \Rightarrow 6K = 16{,}800,\ K = \$2800$

29. $\frac{34{,}720 \text{ ft}}{31 \text{ sec}} = \frac{x}{50 \text{ sec}} \Rightarrow 31x = 1{,}736{,}000$

$x = 56{,}000 \text{ ft}$

31. $\frac{62 \text{ voted}}{100 \text{ reg'd}} = \frac{n}{22{,}550 \text{ reg'd}}$

$100n = 1{,}398{,}100 \Rightarrow n = 13{,}981 \text{ voted}$

33. $\frac{L}{W} = \frac{2 \text{ in.}}{3 \text{ in.}} = \frac{6 \text{ in.}}{W} \Rightarrow 2W = 18$

$W = 9 \text{ in.}$

$\frac{H}{L} = \frac{5 \text{ in.}}{2 \text{ in.}} = \frac{H}{6 \text{ in.}} \Rightarrow 2H = 30$

$H = 15 \text{ in.}$

Cumulative Review Problems

35. $\frac{1}{3} \cdot \frac{1}{4} = \frac{1 \cdot 1}{3 \cdot 4} = \frac{1}{12}$

37. $x^4 \cdot x^5 = x^{4+5} = x^9$

39. $A = 6 \cdot LW = 6 \cdot 36(78) = 16{,}848 \text{ ft}^2$

Putting Your Skills to Work, Problems for Individual Investigation

1. $\frac{3 \text{ rolls}}{1 \text{ person}} = \frac{n}{192 \text{ people}}$

$n = 3(192) = 576 \text{ rolls}$

2. (a) $\frac{2 \text{ lb}}{24 \text{ people}} = \frac{x}{192 \text{ people}}$

$24x = 192(2) = 384 \Rightarrow x = 16 \text{ lb ham}$

(b) $\frac{2 \text{ lb}}{16 \text{ people}} = \frac{x}{192 \text{ people}}$

$16x = 192(2) = 384 \Rightarrow x = 24 \text{ lb turkey}$

(c) $\frac{2 \text{ lb}}{16 \text{ people}} = \frac{x}{192 \text{ people}}$

$16x = 192(2) = 384 \Rightarrow x = 24 \text{ lb cheese}$

3. $\dfrac{1 \text{ basket}}{6 \text{ people}} = \dfrac{x}{192 \text{ people}}$
$6x = 192 \Rightarrow x = 32$ baskets

4. $\dfrac{1 \text{ cake}}{24 \text{ people}} = \dfrac{x}{192 \text{ people}}$
$24x = 192 \Rightarrow x = 8$ cakes

5. $\dfrac{3 \text{ parts nuts}}{2 \text{ parts candy}} = \dfrac{15 \text{ bags of nuts}}{b}$
$3b = 2(15) = 30 \Rightarrow b = 10$ bags candy

6. $\dfrac{6 \text{ yellow}}{9 \text{ white}} = \dfrac{72 \text{ yellow}}{w}$
$6w = 9(72) = 648$
$w = 108$ white daisies

7. $\dfrac{2 \text{ yd white}}{3 \text{ yd yellow}} = \dfrac{w}{30 \text{ yd yellow}}$
$3w = 2(30) = 60 \Rightarrow w = 20$ yd white

8. $\dfrac{4 \text{ cups flour}}{48 \text{ rolls}} = \dfrac{c}{576 \text{ rolls}}$
$48c = 4(576) = 2304 \Rightarrow c = 48$ cups flour

9. $\dfrac{2 \text{ cups milk}}{48 \text{ rolls}} = \dfrac{c}{576 \text{ rolls}}$
$48c = 2(576) = 1152 \Rightarrow c = 24$ cups milk

Putting Your Skills to Work, Problems for Group Investigation and Cooperative Study

1. $\dfrac{\$3}{48 \text{ rolls}} = \dfrac{c}{576} \Rightarrow 48c = 3(576) = 1728$
$c = \$36$

2. $6(16) + 5(24) + 4(24) = \$312$

3. $\dfrac{\$5}{8 \text{ baskets}}(32 \text{ baskets}) = \$20$

4. $\dfrac{\$28}{2 \text{ cakes}}(8 \text{ cakes}) = \$112$

5. (a)
$\dfrac{4 \text{ servings}}{\text{person}}(192 \text{ people}) = 768$ servings
$\dfrac{2 \text{ cups}}{24 \text{ servings}} = \dfrac{c}{768 \text{ servings}}$
$c = 64$ cups

(b) $\dfrac{4 \text{ cups}}{24 \text{ servings}} = \dfrac{c}{768 \text{ servings}}$
$24c = 3072 \Rightarrow c = 128$ cups

(c) $\dfrac{3 \text{ cups}}{24 \text{ servings}} = \dfrac{c}{768 \text{ servings}}$
$c = 96$ cups

6. $\dfrac{24 \text{ servings}}{\$3} = \dfrac{768 \text{ servings}}{c}$
$c = \$96$

7. $15(3) + 10(2) = \$65$

Chapter 4 Review Problems

1. 322,970 ends in 0 and is thus divisible by 2 and 5. The sum of the digits is 23 which is not divisible by 3 and therefore 322,970 is not divisible by 3.

2. 41,592 is even and thus divisible by 2. 41,592 does not end in 0 or 5 and is thus not divisible by 5. The sum of the digits is 21 which is divisible by 3 and thus 41,592 is divisible by 3.

3. 0: neither; 7, 11: prime;
21, 50, 25,51: composite

4. 1: neither; 7, 13, 41: prime; 32, 12, 50, 6: composite

5. $36 = 2 \cdot 2 \cdot 3 \cdot 3 = 2^2 \cdot 3^2$

6. $56 = 2 \cdot 2 \cdot 2 \cdot 7 = 2^3 \cdot 7$

7. $425 = 5 \cdot 5 \cdot 17 = 5^2 \cdot 17$

8. $90 = 2 \cdot 3 \cdot 3 \cdot 5 = 2 \cdot 3^2 \cdot 5$

9. $\frac{1}{0}$ undefined

10. $\frac{0}{6} = 0$

11. $\frac{z}{z} = 1$

12. $\frac{a}{a} = 1$

13. $\frac{7}{20}$

14. $\frac{27}{69} = \frac{\cancel{3} \cdot 9}{\cancel{3} \cdot 23} = \frac{9}{23}$

15. $\frac{42}{5} = 8\frac{2}{5}$

16. $\frac{55}{6} = 9\frac{1}{6}$

17. $\frac{56}{7} = 8$

18. $\frac{42}{7} = 6$

19. $2\frac{1}{3} = \frac{3 \cdot 2 + 1}{3} = \frac{7}{3}$

20. $4\frac{3}{5} = \frac{5 \cdot 4 + 3}{5} = \frac{23}{5}$

21. $10\frac{2}{5} = \frac{5 \cdot 10 + 2}{5} = \frac{52}{5}$

22. $11\frac{1}{4} = \frac{4 \cdot 11 + 1}{4} = \frac{45}{4}$

23. $\frac{1}{9} \cdot \frac{2}{2} = \frac{2}{18}$

24. $\frac{1}{2} \cdot \frac{11}{11} = \frac{11}{22}$

25. $\frac{2}{3} \cdot \frac{9}{9} = \frac{18}{27}$

26. $\frac{3}{4} \cdot \frac{9}{9} = \frac{27}{36}$

27. $\frac{4}{5} \cdot \frac{7x}{7x} = \frac{28x}{35x}$

28. $\frac{6}{11} \cdot \frac{3y}{3y} = \frac{18y}{33y}$

29. $\frac{55}{75} = \frac{\cancel{5} \cdot 11}{\cancel{5} \cdot 15} = \frac{11}{15}$

30. $\frac{48}{54} = \frac{\cancel{6} \cdot 8}{\cancel{6} \cdot 9} = \frac{8}{9}$

31. $\frac{108}{36} = \frac{\cancel{36} \cdot 3}{\cancel{36}} = 3$

32. $\frac{175}{75}=\frac{\cancel{25}\cdot 7}{\cancel{25}\cdot 3}=\frac{7}{3}=2\frac{1}{3}$

33. $\frac{25\cancel{x}}{60\cancel{x}}=\frac{\cancel{5}\cdot 5}{\cancel{5}\cdot 12}=\frac{5}{12}$

34. $\frac{84\cancel{x}}{105\cancel{x}y}=\frac{\cancel{21}\cdot 4}{\cancel{21}\cdot 5y}=\frac{4}{5y}$

35. $\frac{-16}{18}=-\frac{\cancel{2}\cdot 8}{\cancel{2}\cdot 9}=-\frac{8}{9}$

36. $\frac{24}{-36}=-\frac{\cancel{12}\cdot 2}{\cancel{12}\cdot 3}=-\frac{2}{3}$

37. $\frac{y^5}{y^3}=y^{5-3}=y^2$

38. $\frac{3^2}{3^3}=3^{2-3}=3^{-1}=\frac{1}{3}$

39. $\frac{a^2b^4}{a^5b^5}=a^{2-5}b^{4-5}=a^{-3}b^{-1}=\frac{1}{a^3b}$

40. $\frac{x^5y^3}{x^2y^9}=x^{5-2}y^{3-9}=x^3y^{-6}=\frac{x^3}{y^6}$

41. $\frac{2^3x^0}{2^6x^9}=\frac{2^{3-6}\cdot 1}{x^9}=\frac{2^{-3}}{x^9}=\frac{1}{2^3x^9}$

42. $\frac{3^2y^0}{3^3y^6}=\frac{3^{2-3}\cdot 1}{y^6}=\frac{3^{-1}}{y^6}=\frac{1}{3y^6}$

43. $\frac{20x^5}{35x^9}=\frac{\cancel{5}\cdot 4x^{5-9}}{\cancel{5}\cdot 7}=\frac{4x^{-4}}{7}=\frac{4}{7x^4}$

44. $\frac{\overset{3}{\cancel{18}}\,y^6}{\cancel{6}\,y^4}=3y^{6-4}=3y^2$

45. $(y^2)^3=y^{2\cdot 3}=y^6$

46. $(2^4)^2=2^{4\cdot 2}=2^8=256$

47. $\left(\frac{3}{y}\right)^2=\frac{3^2}{y^2}=\frac{9}{y^2}$

48. $\left(\frac{x}{2}\right)^3=\frac{x^3}{2^3}=\frac{x^3}{8}$

49. $\frac{20}{46}=\frac{\cancel{2}\cdot 10}{\cancel{2}\cdot 23}=\frac{10}{23}$

50. $\frac{15}{25}=\frac{\cancel{5}\cdot 3}{\cancel{5}\cdot 5}=\frac{3}{5}$

51. $\frac{35\text{ yd}}{55\text{ yd}}=\frac{\cancel{5}\cdot 7}{\cancel{5}\cdot 11}=\frac{7}{11}$

52. $\frac{24}{28}=\frac{\cancel{4}\cdot 6}{\cancel{4}\cdot 7}=\frac{6}{7}$

53. $\frac{28}{24}=\frac{\cancel{4}\cdot 7}{\cancel{4}\cdot 6}=\frac{7}{6}$

54. $\frac{\$35}{7\text{ washcloths}}=\$5$ per washcloth

55. $\frac{112\text{ in.}}{28\text{ hr}}=4$ in. per hr

56. $\frac{210\text{ mi}}{8\text{ hr}}=26\frac{1}{4}$ mph

57. $\dfrac{10 \text{ fat calories}}{2 \text{ gram protein}}$
$= 5$ fat calories per gram protein

58. $\dfrac{110 \text{ calories}}{24 \text{ gram carbohydrates}}$
$= 4\dfrac{7}{12}$ calories per gram carbohydrates

59. (a) $\dfrac{32 \text{ legal secretaries}}{16 \text{ lawyers}}$
$= 2$ legal secretaries per lawyer

(b) $\dfrac{12 \text{ paralegals}}{4 \text{ lawyers}}$
$= 3$ paralegals per lawyer

(c) $\dfrac{12 \text{ paralegals}}{4 \text{ lawyers}} = \dfrac{p}{60 \text{ lawyers}}$
$p = 180$ paralegals

60. (a) $\dfrac{\$72}{6 \text{ CD's}} = \$12$ per CD
$\dfrac{\$96}{8 \text{ CD's}} = \$12$ per CD
(b) Both have the same unit price.

61. (a) $\dfrac{24{,}000}{31{,}000} = \dfrac{24}{31}$
(b) $\dfrac{\$45{,}000}{30 \text{ days}} = \$1500$ per day

62. $3(70) = 210 \neq 156 = 4(39),\ \dfrac{3}{4} \neq \dfrac{39}{70}$

63. $13(84) = 1092 = 91(12),\ \dfrac{13}{91} = \dfrac{12}{84}$

64. $\dfrac{2}{7} = \dfrac{14}{49}$

65. $\dfrac{\frac{1}{5}}{\frac{1}{9}} = \dfrac{\frac{1}{2}}{\frac{5}{18}}$

66. $\dfrac{2 \text{ in.}}{190 \text{ mi}} = \dfrac{6 \text{ in.}}{570 \text{ mi}}$

67. $\dfrac{234 \text{ mi}}{9 \text{ gal}} = \dfrac{468 \text{ mi}}{18 \text{ gal}}$

68. $3(20) = 60 = 5(12)$, yes

69. $4(46) = 184 \neq 140 = 7(20)$, no

70. $\dfrac{x}{8} = \dfrac{3}{2} \Rightarrow 2x = 8(3) = 24 \Rightarrow x = 12$
check: $\dfrac{12}{8} \stackrel{?}{=} \dfrac{3}{2},\ \dfrac{3}{2} = \dfrac{3}{2}$

71. $\dfrac{14}{x} = \dfrac{7}{3} \Rightarrow 7x = 14(3) = 42 \Rightarrow x = 6$
check: $\dfrac{14}{6} \stackrel{?}{=} \dfrac{7}{3},\ \dfrac{7}{3} = \dfrac{7}{3}$

72. $\dfrac{12}{20} = \dfrac{x}{5} \Rightarrow 20x = 12(5) = 60 \Rightarrow x = 3$
check: $\dfrac{12}{20} \stackrel{?}{=} \dfrac{3}{5},\ \dfrac{3}{5} = \dfrac{3}{5}$

73. $\dfrac{17 \text{ quarts}}{47 \text{ ft}^2} = \dfrac{n \text{ quarts}}{94 \text{ ft}^2} \Rightarrow 47n = 17(94)$
$47n = 1598 \Rightarrow n = 34$

74. $\dfrac{10 \text{ mi}}{2 \text{ hr}} = \dfrac{25 \text{ mi}}{n \text{ hr}} \Rightarrow 10n = 2(25) = 50$
$n = 5$

75. $\dfrac{120 \text{ mi}}{1 \text{ in.}} = \dfrac{x \text{ mi}}{3 \text{ in.}} \Rightarrow x = 120(3) = 360$ mi

76. $\frac{75 \text{ rpm}}{12 \text{ mph}} = \frac{100 \text{ rpm}}{r} \Rightarrow 75r = 1200$

$r = 16$ mi per hr

77. $\frac{4 \text{ ft wide}}{7 \text{ ft long}} = \frac{W}{14 \text{ ft long}} \Rightarrow 7W = 4(14)$

$7W = 56 \Rightarrow W = 8$ ft

78. (a) $\frac{\$400}{\$340} = \frac{\$60}{x} \Rightarrow 400x = 60(340)$

$400x = 20{,}400 \Rightarrow x = \$51$

(b) $\frac{400}{340} = \frac{20}{x} \Rightarrow 400x = 6800 \Rightarrow x = \$17$

Chapter 4 Test

1. 230 is even and therefore divisible by 2. 230 ends in 0 and therefore divisible by 5. Since the sum of the digits, $2+3=5$, is not divisible by 3, 230 is not divisible by 3.

2. (a) $27 = 3^3$: composite
(b) 1 : neither
(c) 19 : prime

3. $84 = 2\cdot 2\cdot 3\cdot 7 = 2^2\cdot 3\cdot 7$

4. $49 = 7\cdot 7 = 7^2$

5. $\frac{0}{4} = 0$

6. $\frac{t}{t} = 1$

7. $\frac{12}{0}$ undefined

8. $\frac{17}{36}$

9. (a) $\frac{16}{12+7+16} = \frac{16}{35}$

(b) $\frac{7+16}{12+7+16} = \frac{23}{35}$

10. $\frac{15}{8}$ improper

11. $\frac{7}{9}$ proper

12. $3\frac{5}{6}$ mixed number

13. $\frac{12}{3} = 4$

14. $\frac{8}{5} = 1\frac{3}{5}$

15. $7\frac{1}{6} = \frac{7\cdot 6+1}{6} = \frac{43}{6}$

16. $\frac{4}{9}\cdot\frac{4}{4} = \frac{16}{36}$

17. $\frac{1}{3}\cdot\frac{4}{4} = \frac{4}{12}$

18. $\frac{7}{8}\cdot\frac{5}{5} = \frac{35}{40}$

19. $\frac{4}{9}\cdot\frac{3y}{3y} = \frac{12y}{27y}$

20. $-\frac{18}{56} = -\frac{\cancel{2}\cdot 9}{\cancel{2}\cdot 28} = -\frac{9}{28}$

21. $\dfrac{16x}{32x^2} = \dfrac{\cancel{2}\cdot\cancel{2}\cdot\cancel{2}\cdot\cancel{2}\cancel{x}}{\cancel{2}\cdot\cancel{2}\cdot\cancel{2}\cdot\cancel{2}\cdot 2\cancel{x}x}$
$= \dfrac{1}{2x}$

22. $\dfrac{5y^3}{20xy} = \dfrac{\cancel{5}\cancel{y}\cdot y^2}{\cancel{5}\cdot 4x\cancel{y}}$
$= \dfrac{y^2}{4x}$

23. $(y^4)^0 = 1$

24. $(x^3)^4 = x^{3\cdot 4} = x^{12}$

25. $\left(\dfrac{x}{3}\right)^2 = \dfrac{x^2}{3^2} = \dfrac{x^2}{9}$

26. $\dfrac{16}{28+16} = \dfrac{16}{44} = \dfrac{\cancel{4}\cdot 4}{\cancel{4}\cdot 11} = \dfrac{4}{11}$

27. $\dfrac{250 \text{ calories}}{5 \text{ grams}} = 50$ calories per gram fat

28. $\dfrac{150 \text{ calories}}{20 \text{ min}} = 7\dfrac{1}{2}$ cal/min

29. $20(13) = 260 = 52(5)$, yes

30. $\dfrac{4}{6} = \dfrac{20}{x} \Rightarrow 4x = 120 \Rightarrow x = 30$

31. $\dfrac{1 \text{ lb}}{1200 \text{ ft}^2} = \dfrac{x}{6000 \text{ ft}^2} \Rightarrow 1200x = 6000$
$x = 5$ lb

Cumulative Test for Chapters 1-4

1. $\$8,400,000 - \$4,575,000 = \$3,825,000$

2. $\$10,000,000 + \$8,666,667 + \$7,885,532$
$= \$26,552,199$

3. (a) $25 - x = 18$
(b) $x = 25 - 18 = 7$

4. $10^3 = 10\cdot 10\cdot 10 = 1000$

5. $8^0 = 1$

6. $12^1 = 12$

7. $(3x)(5x^2) = 3\cdot 5x^{1+2} = 15x^3$

8. $2(x+3) = 2x + 2(3) = 2x + 6$

9. $4(y-2) = 4y - 4(2) = 4y - 8$

10. $a(b+c) = ab + ac$

11. $6n\big|_{n=3} = 6(3) = 18$

12. $-6n\big|_{n=-3} = -6(-3) = 18$

13. 440 is even and thus divisible by 2. Since the sum of the digits, $4+4=8$, is not divisible by 3, 440 is not divisible by 3. 440 ends in 0 and is thus divisible by 5.

14. $110 = 2\cdot 5\cdot 11$

15. $\dfrac{37}{4} = 9\dfrac{1}{4}$

16. $\dfrac{40}{3} = 13\dfrac{1}{3}$

17. $\dfrac{5}{12}\cdot\dfrac{5x}{5x} = \dfrac{25x}{60x}$

18. $\frac{90n^2}{54n} = \frac{\cancel{18}\cdot 5\cancel{n}n}{\cancel{18}\cdot 3\cancel{n}} = \frac{5n}{3}$

19. $\frac{8a^7}{32a^5} = \frac{\cancel{8}\cancel{a^5}\cdot a^2}{\cancel{8}\cdot 4\cancel{a^5}} = \frac{a^2}{4}$

20. $\left(\frac{x}{3}\right)^3 = \frac{x^3}{3^3} = \frac{x^3}{27}$

21. (a) $\frac{44}{16} = \frac{\cancel{4}\cdot 11}{\cancel{4}\cdot 4} = \frac{11}{4}$

(b) $\frac{16}{44} = \frac{\cancel{4}\cdot 4}{\cancel{4}\cdot 11} = \frac{4}{11}$

22. $\frac{310 \text{ mi}}{6 \text{ hr}} = 51\frac{2}{3}$ mph

23. (a) $\frac{\$108}{12 \text{ lessons}} = \$9$ per lesson

$\frac{\$150}{15 \text{ lessons}} = \$10$ per lesson

(b) The 12 lesson special is the better deal.

24. $\frac{150}{4} = \frac{900}{x} \Rightarrow 150x = 3600$

$x = 24$ defective

25. (a) $\frac{200}{350} = \frac{\cancel{50}\cdot 4}{\cancel{50}\cdot 7} = \frac{4}{7}$

(b) $\frac{350-200}{350} = \frac{150}{350} = \frac{\cancel{50}\cdot 3}{\cancel{50}\cdot 7} = \frac{3}{7}$

26. $\frac{225 \text{ words}}{3 \text{ min}} = 75$ words/min

27. $\frac{4 \text{ oz}}{9 \text{ g}} = \frac{12 \text{ oz}}{x} \Rightarrow 4x = 108 \Rightarrow x = 27$ g

Chapter 5

Pretest Chapter 5

1. $\frac{2}{3}\cdot\frac{5}{7}=\frac{2\cdot5}{3\cdot7}=\frac{10}{21}$

2. $\frac{3x^4}{20}\cdot\frac{16}{9x^3}=\frac{\cancel{3}\cdot\cancel{4}\cdot4x^{4-3}}{\cancel{4}\cdot5\cdot\cancel{3}\cdot3}=\frac{4x}{15}$

3. The reciprocal of -9 is $\frac{1}{-9}=-\frac{1}{9}$.

4. $-\frac{5}{8}\div\frac{3}{4}=-\frac{5}{\underset{2}{\cancel{8}}}\cdot\frac{\cancel{4}}{3}=-\frac{5}{6}$

5. $18\div\frac{2}{3}=18\cdot\frac{3}{2}=27$ bags

6. $7,\ 7\cdot2=14,\ 7\cdot3=21,\ 7\cdot4=28$

7. $20=2\cdot2\cdot5,\ 15=3\cdot5$

$\text{LCM}=2^2\cdot3\cdot5=60$

8. $3x=3x,\ 6x^2=3\cdot2x^2,\ 9x=3\cdot3x$

$\text{LCM}=2\cdot3^2x^2=18x^2$

9. $7=7,\ 35=5\cdot7,\ 91=7\cdot13$

$\text{LCM}=5\cdot7\cdot13=455$

10. $7=7,\ 3=3,\ \text{LCD}=3\cdot7=21$

11. $\frac{5}{12}+\frac{3}{12}=\frac{5+3}{12}=\frac{\overset{2}{\cancel{8}}}{\underset{3}{\cancel{12}}}=\frac{2}{3}$

12. $\frac{3}{7x}+\frac{5}{x}\cdot\frac{7}{7}=\frac{3+35}{7x}=\frac{38}{7x}$

13. $\frac{3x}{4}\cdot\frac{3}{3}-\frac{7x}{12}=\frac{9x-7x}{12}=\frac{\overset{1}{\cancel{2}}x}{\underset{6}{\cancel{12}}}=\frac{x}{6}$

14. $\frac{3}{8}\cdot\frac{3}{3}+\frac{1}{6}\cdot\frac{4}{4}=\frac{9+4}{24}=\frac{13}{24}$

15. $1\frac{7}{10}+8\frac{9}{10}=9\frac{16}{10}=10\frac{3}{5}$

16. $25\frac{3}{4}\cdot\frac{2}{2}+13\frac{5}{8}=(25+13)\frac{6+5}{8}$

$=38\frac{11}{8}=39\frac{3}{8}$

17. $6\frac{9}{14}-2\frac{3}{7}\cdot\frac{2}{2}=(6-2)\left(\frac{9}{14}-\frac{6}{14}\right)=4\frac{3}{14}$

18. $3\frac{1}{4}(-2)=\frac{13}{4}(-2)=-\frac{13}{2}=-6\frac{1}{2}$

19. $7\frac{1}{2}\div3\frac{3}{4}=\frac{\cancel{15}}{\underset{1}{\cancel{2}}}\cdot\frac{\overset{2}{\cancel{4}}}{\cancel{15}}=2$

20. $\frac{1}{18}+\frac{2}{3}\cdot\frac{1}{3}\cdot\frac{2}{2}=\frac{1+4}{18}=\frac{5}{18}$

21. $\frac{3}{4}+\frac{1}{2}\div\frac{2}{3}=\frac{3}{4}+\frac{1}{2}\cdot\frac{3}{2}$

$=\frac{3+3}{4}=\frac{6}{4}=\frac{3}{2}$

22. $\frac{\frac{1}{2}}{\frac{3}{8}}=\frac{1}{\underset{1}{\cancel{2}}}\cdot\frac{\overset{4}{\cancel{8}}}{3}=\frac{4}{3}=1\frac{1}{3}$

23. (a) $35\frac{1}{2} - 2\cdot 2\frac{1}{2} = \frac{71}{2} - \frac{10}{2} = \frac{61}{2} = 30\frac{1}{2}$

$24 - 2\cdot 2\frac{1}{2} = \frac{48}{2} - \frac{10}{2} = \frac{38}{2} = 19$

dimensions: $30\frac{1}{2}$ ft by 19 ft

(b) $\left(2\cdot 30\frac{1}{2} + 2\cdot 19\right)\cdot 2\frac{1}{4}$

$= \left(\cancel{2}\cdot\frac{61}{\cancel{2}} + 38\right)\cdot\frac{9}{4} = \frac{891}{4} = \$222\frac{3}{4}$

24. $\frac{x}{-6} = 14 \Rightarrow x = (-6)(14) = -84$

check: $\frac{-84}{-6} \stackrel{?}{=} 14,\ 14 = 14$

25. $-\frac{5}{7}x = 10 \Rightarrow x = -\frac{7}{5}\cdot 10 = -14$

check: $-\frac{5}{7}(-14) \stackrel{?}{=} 10,\ 10 = 10$

26. $\frac{7}{8}x = 56 \Rightarrow x = \frac{8}{7}\cdot 56 = 64$

check: $\frac{7}{8}\cdot 64 \stackrel{?}{=} 56,\ 56 = 56$

5.1 Exercises

1. $\frac{1}{3}\cdot\frac{1}{5} = \frac{1\cdot 1}{3\cdot 5} = \frac{1}{15}$

3. $\frac{5}{\cancel{21}_{3}}\cdot\frac{\cancel{7}^{1}}{8} = \frac{5\cdot 1}{3\cdot 8} = \frac{5}{24}$

5. $\frac{\cancel{7}^{1}}{\cancel{12}_{3}}\cdot\frac{\cancel{8}^{2}}{\cancel{28}_{4}} = \frac{1\cdot\cancel{2}^{1}}{3\cdot\cancel{4}_{2}} = \frac{1}{6}$

7. $\frac{\cancel{3}^{1}}{\cancel{42}_{7}}\cdot\frac{\cancel{6}^{1}}{\cancel{15}_{5}} = \frac{1\cdot 1}{7\cdot 5} = \frac{1}{35}$

9. $\frac{\cancel{-3}^{1}}{\cancel{8}_{1}}\cdot\frac{\cancel{32}^{4}}{\cancel{-6}_{2}} = \frac{\cancel{4}^{2}}{\cancel{2}_{1}} = 2$

11. $\frac{\cancel{16}^{4}}{11}\cdot\frac{-18}{\cancel{36}_{9}} = -\frac{4\cdot\cancel{18}^{2}}{11\cdot\cancel{9}_{1}} = -\frac{8}{11}$

13. $\frac{6}{\cancel{35}_{7}}\cdot\cancel{5}^{1} = \frac{6}{7}$

15. $-\cancel{14}^{1}\cdot\frac{1}{\cancel{28}_{2}} = -\frac{1}{2}$

17. $\frac{-\cancel{2}^{1}}{\cancel{63}_{9}}\cdot\left(\frac{-\cancel{14}^{2}}{\cancel{18}_{9}}\right) = \frac{2}{81}$

19. $\frac{1}{\cancel{5}_{1}}\cdot\cancel{25}^{5} = 5$

21. $\frac{\cancel{2}^{1}x^2}{\cancel{9}_{3}}\cdot\frac{\cancel{3}^{1}x^2}{\cancel{8}_{4}} = \frac{x^4}{12}$

23. $\frac{\cancel{6}^{1}\cancel{x}}{\cancel{25}_{5}}\cdot\frac{\cancel{15}^{3}}{\cancel{12}_{2}x^{\cancel{2}}} = \frac{3}{10x}$

25. $\frac{-\cancel{3}^{1}y^{\cancel{3}}}{\cancel{20}_{5}}\cdot\frac{\cancel{12}^{3}}{\cancel{21}_{7}\cancel{y^2}} = -\frac{3y}{35}$

27. $\frac{\overset{1}{\cancel{3}}x^2}{\underset{5}{\cancel{15}}}\cdot\frac{\overset{9}{\cancel{18}}x^3}{\underset{10}{\cancel{20}}}=\frac{9x^5}{50}$

29. $\frac{6x^4}{\underset{1}{\cancel{7}}}\cdot\overset{2}{\cancel{14}}x^3=12x^7$

31. $A=\frac{1}{2}bh=\frac{1}{2}(14)(9)=63\text{ m}^2$

33. $A=\frac{1}{2}bh=\frac{1}{2}(21)(40)=420\text{ in.}^2$

35. The reciprocal of $\frac{1}{8}$ is $\frac{8}{1}=8$

37. The reciprocal of 8 is $\frac{1}{8}$

39. The reciprocal of $\frac{2}{-5}$ is $\frac{-5}{2}=-\frac{5}{2}$

41. The reciprocal of $\frac{-x}{y}$ is $\frac{y}{-x}=-\frac{y}{x}$

43. The reciprocal of -6 is $\frac{1}{-6}=-\frac{1}{6}$

45. $\frac{6}{14}\div\frac{3}{8}=\frac{\overset{\cancel{2}}{\cancel{6}}}{\underset{7}{\cancel{14}}}\cdot\frac{8}{\underset{1}{\cancel{3}}}=\frac{8}{7}$

47. $\frac{7}{24}\div\frac{9}{8}=\frac{7}{\underset{3}{\cancel{24}}}\cdot\frac{\cancel{8}}{9}=\frac{7}{27}$

49. $\frac{-1}{12}\div\frac{3}{4}=-\frac{1}{\underset{3}{\cancel{12}}}\cdot\frac{\cancel{4}}{3}=-\frac{1}{9}$

51. $\frac{-7}{24}\div\frac{9}{-8}=\frac{7}{\underset{3}{\cancel{24}}}\cdot\frac{\cancel{8}}{9}=\frac{7}{27}$

53. $15\div\frac{3}{7}=\overset{5}{\cancel{15}}\cdot\frac{7}{\cancel{3}}=35$

55. $\frac{7}{22}\div 14=\frac{\cancel{7}}{22}\cdot\frac{1}{\underset{2}{\cancel{14}}}=\frac{1}{44}$

57. $\frac{-7}{15}\div\frac{-5}{25}=\frac{7}{\underset{3}{\cancel{15}}}\cdot\frac{\overset{\cancel{5}}{\cancel{25}}}{\cancel{5}}=\frac{7}{3}$

59. $\frac{5}{14}\div\frac{2}{21}\div\frac{\overset{5}{\cancel{15}}}{-\cancel{3}}=-\frac{\cancel{5}}{\underset{2}{\cancel{14}}}\cdot\frac{\overset{3}{\cancel{21}}}{2}\cdot\frac{1}{\cancel{5}}=-\frac{3}{4}$

61. $\frac{8x^6}{15}\div\frac{16x^2}{5}=\frac{\cancel{8}x^{\cancel{6}4}}{\underset{3}{\cancel{15}}}\cdot\frac{\cancel{5}}{\underset{2}{\cancel{16}}\cancel{x^2}}=\frac{x^4}{6}$

63. $\frac{7x^4}{12}\div\frac{28}{36x^2}=\frac{\cancel{7}x^4}{\cancel{12}}\cdot\frac{\overset{3}{\cancel{36}}x^2}{\underset{4}{\cancel{28}}}=\frac{3x^6}{4}$

65. $\frac{3}{12}\cdot 3300=\$825$

67. $12\div\frac{3}{4}=\overset{4}{\cancel{12}}\cdot\frac{4}{\cancel{3}}=16$

69. $32\text{ laps}\cdot\frac{\frac{1}{4}\text{ mi}}{\text{laps}}=\frac{32}{4}\text{ mi}=8\text{ mi}$

71. $\frac{120 \text{ qt}}{\text{vat}} \cdot \frac{\text{bottle}}{\frac{3}{4} \text{ qt}} = 120 \cdot \frac{4}{3} = 160 \frac{\text{bottles}}{\text{vat}}$

73. (a) $9600 \cdot \frac{1}{6} = 1600$ students from CA

(b) $9600 \cdot \frac{1}{5} = 1920$ students from TX

75. $\frac{\cancel{3}}{4} \cdot \frac{x}{\cancel{27}_{9}} = \frac{4}{\cancel{9}} \Rightarrow x = 4 \cdot 4 = 16$

Cumulative Review Problems

77. $\frac{2}{3} \cdot \frac{5}{5} = \frac{10}{15}$

79. $\frac{9}{10} \cdot \frac{5}{5} = \frac{45}{50}$

81. $3(4) + 2(6) = 24 > 17$, no

5.2 Exercises

1. Because $3 \cdot 4 = 12$ and there is no whole number that we can multiply by 5 to get 12.

3. (a) 6, 12, 18, 24; 8, 16, 24, 32
(b) 24

5. (a) 2, 4, 6, 8, 10; 5, 10, 15, 20, 25
(b) 10

7. (a) $12x,\ 24x,\ 36x,\ 48x$
$18x,\ 36x,\ 54x,\ 72x$
(b) $36x$

9. $9 = 3 \cdot 3,\ 18 = 2 \cdot 3 \cdot 3,\ \text{LCM} = 2 \cdot 3^2 = 18$

11. $16 = 2 \cdot 2 \cdot 2 \cdot 2,\ 28 = 2 \cdot 2 \cdot 7$
$\text{LCM} = 2^4 \cdot 7 = 112$

13. $15 = 3 \cdot 5,\ 21 = 3 \cdot 7,\ \text{LCM} = 3 \cdot 5 \cdot 7 = 105$

15. $40 = 2 \cdot 2 \cdot 2 \cdot 5,\ 60 = 2 \cdot 2 \cdot 3 \cdot 5$
$\text{LCM} = 2^3 \cdot 3 \cdot 5 = 120$

17. $5 = 5,\ 8 = 2^3,\ 12 = 2^2 \cdot 3$
$\text{LCM} = 2^3 \cdot 3 \cdot 5 = 120$

19. $7 = 7,\ 14 = 2 \cdot 7,\ 35 = 5 \cdot 7$
$\text{LCM} = 2 \cdot 5 \cdot 7 = 70$

21. $4x = 2^2 x,\ 7x = 7x,\ \text{LCM} = 4 \cdot 7x = 28x$

23. $21a = 3 \cdot 7a,\ 81a = 3^4 a$
$\text{LCM} = 3^4 \cdot 7a = 567a$

25. $18x = 2 \cdot 3^2 x,\ 45x^2 = 3^2 \cdot 5x^2$
$\text{LCM} = 2 \cdot 3^2 \cdot 5x^2 = 90x^2$

27. $22x^2 = 2 \cdot 11x^2,\ 4x^3 = 2^2 x^3$
$\text{LCM} = 2^2 \cdot 11x^3 = 44x^3$

29. $12x = 2^2 \cdot 3x,\ 26 = 2 \cdot 13$
$156x^2 = 2^2 \cdot 3 \cdot 13x^2$
$\text{LCM} = 2^2 \cdot 3 \cdot 13x^2 = 156x^2$

31. 2, 3

33. 7, x

35. $\frac{1 \text{ lap}}{4 \text{ min}} \cdot t = \frac{1 \text{ lap}}{6 \text{ min}} \cdot t + 1 \text{ lap}$
$\left(\frac{1}{4} - \frac{1}{6}\right)t = \frac{t}{12} = 1 \Rightarrow t = 12 \text{ min}$

37. $\frac{1 \text{ carton}}{6 \text{ min}} \cdot t = \frac{1 \text{ carton}}{8 \text{ min}} \cdot t + 1 \text{ carton}$

$\left(\frac{1}{6} - \frac{1}{8}\right) \cdot t = \frac{t}{24} = 1 \Rightarrow t = 24 \text{ min}$

39. There 35 min and 45 min between events on the two fields, respectively.
$35 = 5 \cdot 7, 45 = 5 \cdot 9, \text{LCM} = 5 \cdot 7 \cdot 9 = 315$

$315 \text{ min } = 5\frac{1}{4} \text{ hr} \rightarrow 1{:}15 \text{ P.M.}$

41. (a) $660 = 2 \cdot 2 \cdot 3 \cdot 5 \cdot 11 = 2^2 \cdot 3 \cdot 5 \cdot 11$
$140 = 2 \cdot 2 \cdot 5 \cdot 7 = 2^2 \cdot 5 \cdot 7$
(b) $\text{LCM} = 2^2 \cdot 3 \cdot 5 \cdot 7 \cdot 11 = 4620$

43. $2x^3 = 2x^3$
$8xy^2 = 2^3 xy^2$
$10x^2y = 2 \cdot 5x^2 y$
$\text{LCM} = 2^3 \cdot 5 \cdot x^3 \cdot y^2$
$= 40x^3y^2$

45. $2z^2 = 2z^2$
$5xyz = 5xyz$
$15xy = 3 \cdot 5xy$
$\text{LCM} = 2 \cdot 3 \cdot 5xyz^2$
$= 30xyz^2$

Cumulative Review Problems

47. $\frac{15}{2} = 7\frac{1}{2}$

49. $2 + 6(-1) \div 3 = 2 + (-6) \div 3$
$= 2 + (-2)$
$= 0$

51. $-11 + 14 + 7 + (-16) = -6$
The market fell by 6 points.

5.3 Exercises

1. When we add two fractions with the same denominator, we add the <u>numerators</u>, and the <u>denominator</u> stays the same.

3. $\frac{4}{21} + \frac{7}{21} = \frac{4+7}{21} = \frac{11}{21}$

5. $\frac{6}{17} - \frac{3}{17} = \frac{6-3}{17} = \frac{3}{17}$

7. $\frac{-13}{28} + \left(\frac{-11}{28}\right) = \frac{-13+(-11)}{28} = \frac{-24}{28} = \frac{-6}{7}$

9. $\frac{-31}{51} + \frac{11}{51} = \frac{-31+11}{51} = \frac{-20}{51}$

11. $\frac{7}{x} - \frac{5}{x} = \frac{7-5}{x} = \frac{2}{x}$

13. $\frac{31}{a} + \frac{8}{a} = \frac{31+8}{a} = \frac{39}{a}$

15. $\frac{x}{7} - \frac{5}{7} = \frac{x-5}{7}$

17. $\frac{y}{9} + \frac{14}{9} = \frac{y+14}{9}$

19. $4 = 2^2,\ 7 = 7,\ \text{LCD } = 2^2 \cdot 7 = 28$

21. $12 = 2^2 \cdot 3,\ 15 = 3 \cdot 5,\ \text{LCD} = 2^2 \cdot 3 \cdot 5 = 60$

23. $5 = 5,\ 25 = 5^2,\ \text{LCD} = 5^2 = 25$

25. $9 = 3^2,\ 21 = 3 \cdot 7,\ \text{LCD} = 3^2 \cdot 7 = 63$

27. $\frac{1}{4}\cdot\frac{15}{15}=\frac{15}{60}$

29. $\frac{5}{6}\cdot\frac{10}{10}=\frac{50}{60}$

31. $\frac{1}{6}\cdot\frac{12}{12}=\frac{12}{72}$

33. $\frac{5}{9}\cdot\frac{8}{8}=\frac{40}{72}$

35. $\frac{11}{15}\cdot\frac{3}{3}-\frac{31}{45}=\frac{33-31}{45}=\frac{2}{45}$

37. $\frac{16}{24}-\frac{1}{6}\cdot\frac{4}{4}=\frac{16-4}{24}=\frac{12}{24}=\frac{1}{2}$

39. $\frac{3}{8}\cdot\frac{7}{7}+\frac{4}{7}\cdot\frac{8}{8}=\frac{21+32}{56}=\frac{53}{56}$

41. $\frac{5}{6}\cdot\frac{5}{5}+\frac{4}{5}\cdot\frac{6}{6}=\frac{25+24}{30}=\frac{49}{30}=1\frac{19}{30}$

43. $\frac{-3}{4}\cdot\frac{5}{5}+\frac{1}{5}\cdot\frac{4}{4}=\frac{-15+4}{20}=\frac{-11}{20}$

45. $\frac{-2}{13}\cdot\frac{2}{2}+\frac{7}{26}=\frac{-4+7}{26}=\frac{3}{26}$

47. $\frac{-3}{14}\cdot\frac{2}{2}+\left(\frac{-9}{28}\right)=\frac{-6+(-9)}{28}=\frac{-15}{28}$

49. $\frac{3}{16}\cdot\frac{5}{5}+\left(\frac{-9}{20}\right)\cdot\frac{4}{4}=\frac{15+(-36)}{80}=\frac{-21}{80}$

51. $\frac{7}{10}\cdot\frac{10}{10}-\frac{13}{100}=\frac{70-13}{100}=\frac{57}{100}$

53. $\frac{9}{8}\cdot\frac{4}{4}-\frac{5}{32}=\frac{36-5}{32}=\frac{31}{32}$

55. $\frac{5}{2x}+\frac{8}{x}\cdot\frac{2}{2}=\frac{5+16}{2x}=\frac{21}{2x}$

57. $\frac{2}{7x}+\frac{3}{x}\cdot\frac{7}{7}=\frac{2+21}{7x}=\frac{23}{7x}$

59. $\frac{3}{2x}\cdot\frac{3}{3}+\frac{5}{6x}\cdot=\frac{9+5}{6x}=\frac{14}{6x}=\frac{7}{3x}$

61. $\frac{3}{x}\cdot\frac{y}{y}+\frac{4}{y}\cdot\frac{x}{x}=\frac{3y+4x}{xy}$

63. $\frac{2}{x}\cdot\frac{y}{y}-\frac{7}{y}\cdot\frac{x}{x}=\frac{2y-7x}{xy}$

65. $\frac{9}{y}\cdot\frac{x}{x}+\frac{1}{x}\cdot\frac{y}{y}=\frac{9x+y}{xy}$

67. $\frac{4x}{15}+\frac{3x}{5}\cdot\frac{3}{3}=\frac{4x+9x}{15}=\frac{13x}{15}$

69. $\frac{8x}{10}\cdot\frac{2}{2}-\frac{7x}{20}=\frac{16x-7x}{20}=\frac{9x}{20}$

71. $\frac{x}{3}\cdot\frac{4}{4}+\frac{6x}{12}=\frac{4x+6x}{12}=\frac{10x}{12}=\frac{5x}{6}$

73. $\frac{2}{3}\cdot\frac{4}{4}+\frac{1}{4}\cdot\frac{3}{3}=\frac{8+3}{12}=\frac{11}{12}$

75. (a) $\frac{3}{4}+\frac{1}{2}\cdot\frac{2}{2}=\frac{3+2}{4}=\frac{5}{4}=1\frac{1}{4}$ cups sugar

(b) $\frac{3}{4}-\frac{1}{2}\cdot\frac{2}{2}=\frac{3-2}{4}=\frac{1}{4}$ cup

77. $\frac{1}{3}\cdot\frac{4}{4}-\frac{1}{4}\cdot\frac{3}{3}=\frac{4-3}{12}=\frac{1}{12}$

79. $\frac{5}{30}\cdot\frac{4}{4}+\frac{3}{40}\cdot\frac{3}{3}+\frac{1}{8}\cdot\frac{15}{15}=\frac{20+9+15}{120}$
$=\frac{44}{120}=\frac{\not{4}\cdot 11}{\not{4}\cdot 30}=\frac{11}{30}$

81. $\frac{1}{3}\cdot\frac{4}{4}+\frac{1}{12}-\frac{1}{6}\cdot\frac{2}{2}=\frac{4+1-2}{12}=\frac{3}{12}=\frac{1}{4}$

83. $\frac{x}{7}+\frac{1}{7}=\frac{3}{7}\Rightarrow\frac{x+1}{7}=\frac{3}{7}$
$x+1=3\Rightarrow x=2$

85. $\frac{x}{4}-\frac{1}{4}=\frac{1}{4}\Rightarrow\frac{x-1}{4}=\frac{1}{4}$
$x-1=1\Rightarrow x=2$

Cumulative Review Problems

87. $x-3=70\Rightarrow x=70+3=73$

89. $x-9=29\Rightarrow x=29+9=38$

91. $3033-1295-469-387-287=\$595$

Putting Your Skills to Work, Problems for Individual Investigation

1. $\frac{\text{1 beat}}{\text{quarter-note}}\cdot n$ quarter-note = 4 beats
$n=4$. Four quarter-notes

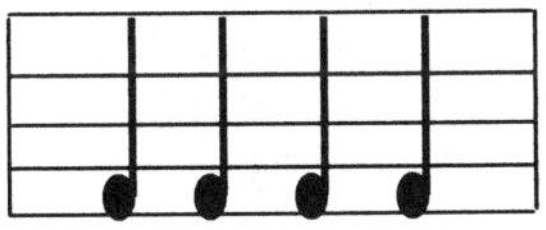

2. $\frac{\text{2 beat}}{\text{half-note}}\cdot n$ half-note = 4 beats, $n=2$
Two half-notes

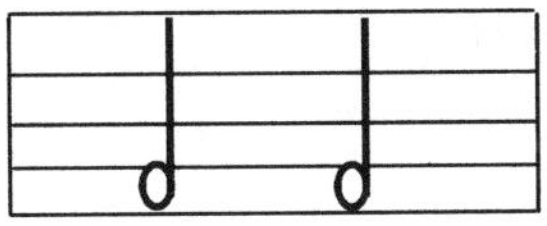

3. $\frac{\text{2 beats}}{\text{half-note}}\cdot 1+\frac{\text{1 beat}}{\text{quarter-note}}\cdot 2=4$ beats
One half-note and two quarter-notes

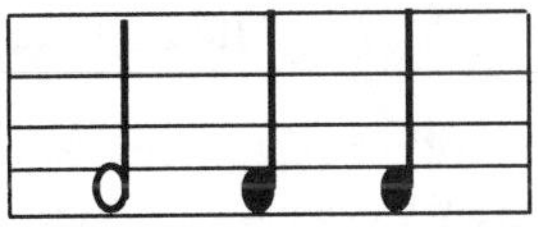

4. Four, since this measure contains two beats, i.e., two quarter-notes with 2(2) eighth-notes.

5. Four. Since the half-note takes up two beats out of six, there are four beats left for quarter-notes.

6. 1 half-note, 2 eighth-notes, and 1 quarter-note:

1 half-note and 4 eighth-notes:

1 half-note and two quarter-notes:

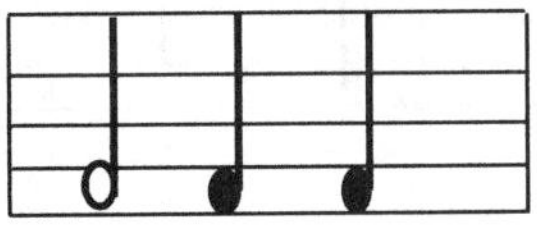

2 half-notes:

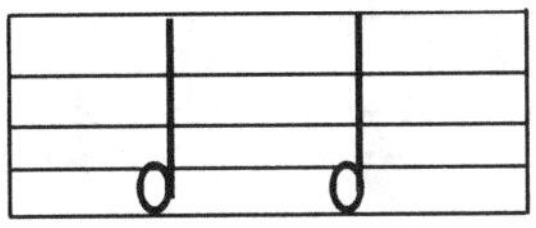

Putting Your Skills to Work, Problems for Group Investigation and Cooperative Study

1. Think of $\frac{3}{4}$ as $3/4$. The numerator tells us how many quarter-notes can occur in a measure of 3 over 4. Since there are 3 beats for the measure and the quarter-note gets one beat. there are 3 quarter-notes in the measure.

2. Think $3/4 = 6/8$. The numerator 6 indicates there are 6 eighth-notes.

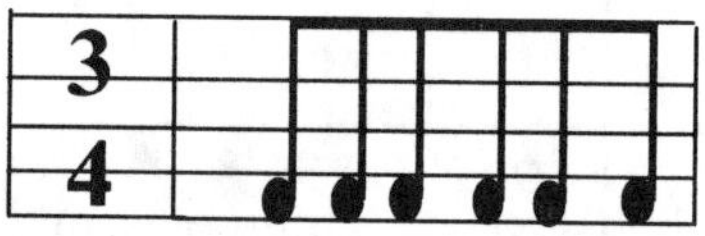

3. Think $4/4 = 8/8$.There are 8 eighth-notes.

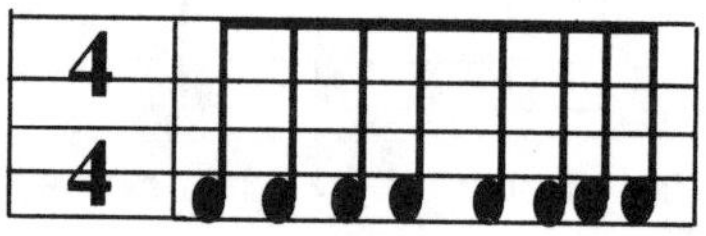

4. Two, think $4/4 = 2/2$. The numerator, 2, indicates that there are 2 half-notes.

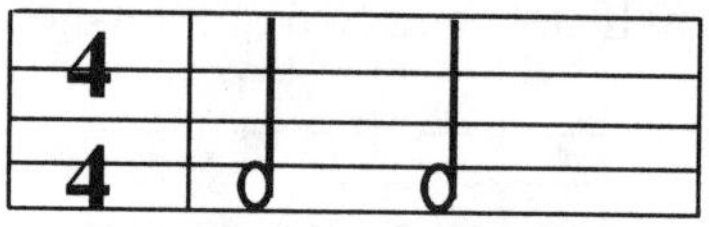

5. 1 half-note and 2 eighth-notes:

5. 1 half-note and 1 quarter-note:

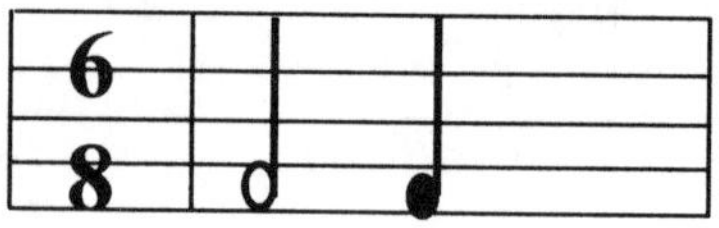

6. Two dotted half-notes would fill the measure:

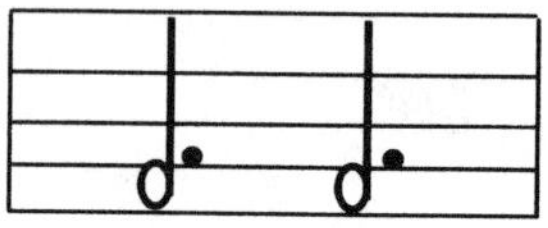

One dotted half-note and three quarter-notes:

5.4 Exercises

1. (a) Marcy did not change the mixed numbers to improper fractions before she multiplied.

(b) $2\frac{2}{3}\cdot 3\frac{4}{5}=\frac{8}{3}\cdot\frac{19}{5}=\frac{152}{15}=10\frac{2}{15}$

3.
$$\begin{array}{r} 12\frac{3}{7} \\ +15\frac{2}{7} \\ \hline 27\frac{5}{7} \end{array}$$

5.
$$\begin{array}{r} 5\frac{5}{8} \\ +11\frac{1}{8} \\ \hline 16\frac{6}{8}=16\frac{3}{4} \end{array}$$

7. $11\frac{2}{3}\cdot\frac{4}{4}=11\frac{8}{12}$
$+7\frac{1}{4}\cdot\frac{3}{3}=\ 7\frac{3}{12}$
$18\frac{11}{12}$

9. $7\frac{4}{5}\cdot\frac{2}{2}=\ 7\frac{8}{10}$
$-2\frac{1}{10}=-2\frac{1}{10}$
$5\frac{7}{10}$

11. $9\frac{2}{3}\cdot\frac{2}{2}=9\frac{4}{6}$
$-6\frac{1}{6}\quad -6\frac{1}{6}$
$3\frac{3}{6}=3\frac{1}{2}$

13. $1\frac{1}{6}\cdot\frac{4}{4}=\ 1\frac{4}{24}$
$+\frac{3}{8}\cdot\frac{3}{3}=+\frac{9}{24}$
$1\frac{13}{24}$

15. $25\frac{2}{3}\cdot\frac{7}{7}=25\frac{14}{21}$
$-6\frac{1}{7}\cdot\frac{3}{3}=-6\frac{3}{21}$
$19\frac{11}{21}$

17. $14\frac{7}{9}\quad 14\frac{7}{9}$
$+6\frac{1}{3}\cdot\frac{3}{3}=+6\frac{3}{9}$
$20\frac{10}{9}=21\frac{1}{9}$

19. $11\frac{1}{5}\cdot\frac{5}{5}=11\frac{5}{25}=10\frac{30}{25}$
$-6\frac{11}{25}\qquad -6\frac{11}{25}$
$4\frac{19}{25}$

21. $10\frac{5}{12}\cdot\frac{5}{5}=10\frac{25}{60}=\ 9\frac{85}{60}$
$-3\frac{9}{10}\cdot\frac{6}{6}=-3\frac{54}{60}=-3\frac{54}{60}$
$6\frac{31}{60}$

23. $6\frac{5}{6}\cdot\frac{4}{4}=\ 6\frac{20}{24}$
$+4\frac{3}{8}\cdot\frac{3}{3}=\ 4\frac{9}{24}$
$10\frac{29}{24}=11\frac{5}{24}$

25. $8\frac{1}{4}\cdot\frac{3}{3}=\ 8\frac{3}{12}$
$+3\frac{5}{6}\cdot\frac{2}{2}=+3\frac{10}{12}$
$11\frac{13}{12}=12\frac{1}{12}$

27. $9 = 8\frac{4}{4}$

$$\begin{array}{rr} & 8\frac{4}{4} \\ -2\frac{1}{4} & -2\frac{1}{4} \\ \hline & 6\frac{3}{4} \end{array}$$

29. $32 = 31\frac{9}{9}$

$$\begin{array}{rr} & 31\frac{9}{9} \\ -1\frac{2}{9} & -1\frac{2}{9} \\ \hline & 30\frac{7}{9} \end{array}$$

31. $2\frac{1}{5}\cdot 1\frac{2}{3} = \frac{11}{\cancel{5}}\cdot\frac{\cancel{5}}{3} = \frac{11}{3} = 3\frac{2}{3}$

33. $4\frac{1}{3}\cdot 2\frac{1}{4} = \frac{13}{\cancel{3}}\cdot\frac{\overset{3}{\cancel{9}}}{4} = \frac{39}{4} = 9\frac{3}{4}$

35. $2\frac{3}{4} \div 1\frac{7}{10} = \frac{11}{\underset{2}{\cancel{4}}}\cdot\frac{\overset{5}{\cancel{10}}}{17} = \frac{55}{34} = 1\frac{21}{34}$

37. $4\frac{1}{2} \div 2\frac{1}{4} = \frac{\cancel{9}}{\cancel{2}}\cdot\frac{\overset{2}{\cancel{4}}}{\cancel{9}} = 2$

39. $1\frac{1}{4}\cdot 3\frac{2}{3} = \frac{5}{4}\cdot\frac{11}{3} = \frac{55}{12} = 4\frac{7}{12}$

41. $-\frac{3}{4}\cdot 9\frac{5}{7} = -\frac{3}{\cancel{4}}\cdot\frac{\overset{17}{\cancel{68}}}{7} = -\frac{51}{7} = -7\frac{2}{7}$

43. $3\frac{1}{4} \div \frac{3}{8} = \frac{13}{\cancel{4}}\cdot\frac{\overset{2}{\cancel{8}}}{3} = \frac{26}{3} = 8\frac{2}{3}$

45. $-\frac{1}{4} \div 1\frac{7}{8} = -\frac{1}{\cancel{4}}\cdot\frac{\overset{2}{\cancel{8}}}{15} = -\frac{2}{15}$

47. $-5 \div \frac{1}{4} = -5\cdot\frac{4}{1} = -20$

49. $7\frac{1}{2} \div \frac{6}{6} = 7\frac{1}{2} \div 1 = 7\frac{1}{2}$

51. $26 \div 6\frac{1}{2} = 26\cdot\frac{2}{13} = 4$ pieces

53. $2\frac{1}{4}\cdot 2 = \frac{9}{4}\cdot 2 = \frac{9}{2} = 4\frac{1}{2}$ cups flour

55. $2\frac{1}{2}\cdot 4 = \frac{5}{2}\cdot 4 = 10$ cups chocolate chips

57. $9\frac{1}{8} + 12\frac{1}{3} + 17\frac{1}{6} = 9\frac{3}{24} + 12\frac{8}{24} + 17\frac{4}{24}$

$= 38\frac{15}{24}$

$= 38\frac{5}{8}$

Cumulative Review Problems

59. $2 + 9\cdot 8 = 2 + 72 = 74$

61. $\frac{(5+7)}{(2\cdot 3)} = \frac{12}{6} = 2$

63. $19(7) + 28(2) - 2(3) - 1(5) = \$178$ Jan.
$26(7) + 36(2) - 3(3) - 2(5) = \$235$ Feb.
$26(7) + 31(2) - 3(3) - 2(5) = \$225$ Mar.
$178 + 235 + 225 = \$638$ total

5.5 Exercises

1. Multiply

3. $\frac{3}{5}-\frac{1}{3}\div\frac{5}{6}=\frac{3}{5}-\frac{1}{\cancel{3}}\cdot\frac{\overset{2}{\cancel{6}}}{5}=\frac{3}{5}-\frac{2}{5}=\frac{1}{5}$

5. $\frac{3}{4}\cdot\frac{5}{5}+\frac{1}{4}\cdot\frac{3}{5}=\frac{15+3}{20}=\frac{18}{20}=\frac{\cancel{2}\cdot 9}{\cancel{2}\cdot 10}=\frac{9}{10}$

7. $\frac{5}{7}\cdot\frac{1}{3}\div\frac{2}{7}=\frac{5}{\underset{3}{\cancel{21}}}\cdot\frac{\cancel{7}}{2}=\frac{5}{6}$

9. $\frac{1}{2}\cdot\frac{3}{3}-\frac{2}{6}+\frac{1}{3}\cdot\frac{2}{2}=\frac{3-2+2}{6}=\frac{3}{6}=\frac{1}{2}$

11. $\frac{5}{6}\cdot\frac{1}{2}+\frac{2}{3}\div\frac{4}{3}=\frac{5}{12}+\frac{2}{3}\cdot\frac{3}{4}=\frac{5}{12}+\frac{1}{2}\cdot\frac{6}{6}$

$=\frac{5+6}{12}=\frac{11}{12}$

13. $\frac{\cancel{2}}{9}\cdot\frac{1}{\underset{2}{\cancel{4}}}+\left(\frac{2}{3}\div\frac{6}{7}\right)=\frac{1}{18}+\left(\frac{\cancel{2}}{3}\cdot\frac{7}{\underset{3}{\cancel{6}}}\right)$

$=\frac{1}{18}+\frac{7}{9}\cdot\frac{2}{2}=\frac{1+14}{18}=\frac{\overset{5}{\cancel{15}}}{\underset{6}{\cancel{18}}}=\frac{5}{6}$

15. $\frac{3}{4}\cdot\frac{1}{4}+\left(\frac{3}{4}\right)^2=\frac{3}{16}+\frac{9}{16}=\frac{12}{16}=\frac{3}{4}$

17. $\left(-\frac{2}{5}\right)\cdot\left(\frac{1}{4}\right)^2=-\frac{2}{5}\cdot\frac{1}{16}=-\frac{1}{40}$

19. $x-\frac{2}{5}\div\frac{4}{15}\Big|_{x=\frac{7}{2}}=\frac{7}{2}\cdot\frac{5}{5}-\frac{\cancel{2}}{5}\cdot\frac{15}{\underset{2}{\cancel{4}}}=\frac{35-15}{10}$

$=\frac{20}{10}=2$

21. $-\frac{3}{8}\cdot\frac{16}{21}+x\Big|_{x=-\frac{4}{7}}=-\frac{\cancel{3}}{\cancel{8}}\cdot\frac{\overset{2}{\cancel{16}}}{\underset{7}{\cancel{21}}}+\left(-\frac{4}{7}\right)$

$=-\frac{2}{7}-\frac{4}{7}=-\frac{6}{7}$

23. $\frac{\frac{2}{3}}{\frac{8}{9}}=\frac{\cancel{2}}{\cancel{3}}\cdot\frac{\overset{3}{\cancel{9}}}{\underset{4}{\cancel{8}}}=\frac{3}{4}$

25. $\frac{\frac{4}{7}}{\frac{12}{21}}=\frac{\cancel{4}}{\cancel{7}}\cdot\frac{\overset{3}{\cancel{21}}}{\underset{3}{\cancel{12}}}=\frac{3}{3}=1$

27. $\frac{\frac{6}{7}}{\frac{9}{14}}=\frac{\overset{2}{\cancel{6}}}{\cancel{7}}\cdot\frac{\overset{2}{\cancel{14}}}{\underset{3}{\cancel{9}}}=\frac{4}{3}$

29. $\frac{\frac{x^2}{3}}{\frac{x}{6}}=\frac{x^{\cancel{2}}}{\cancel{3}}\cdot\frac{\overset{2}{\cancel{6}}}{\cancel{x}}=2x$

31. $\frac{\frac{x}{4}}{\frac{x^2}{12}}=\frac{\cancel{x}}{\cancel{4}}\cdot\frac{\overset{3}{\cancel{12}}}{x^{\cancel{2}}}=\frac{3}{x}$

33. $\frac{\frac{x}{8}}{\frac{x^2}{16}}=\frac{\cancel{x}}{\cancel{8}}\cdot\frac{\overset{2}{\cancel{16}}}{x^{\cancel{2}}}=\frac{2}{x}$

35. $\frac{\frac{1}{2}+\frac{3}{4}}{\frac{4}{5}+\frac{1}{10}}\cdot\frac{20}{20}=\frac{10+15}{16+2}=\frac{25}{18}$

37. $\dfrac{\frac{1}{9}+\frac{2}{3}}{\frac{3}{2}+\frac{1}{3}}\cdot\dfrac{18}{18}=\dfrac{2+12}{27+6}=\dfrac{14}{33}$

39. $\dfrac{\frac{4}{25}-\frac{3}{50}}{\frac{3}{10}+\frac{5}{20}}\cdot\dfrac{100}{100}=\dfrac{16-6}{30+25}=\dfrac{10}{55}=\dfrac{2}{11}$

41. $\dfrac{\frac{3}{7}\cdot\frac{3}{3}-\frac{5}{21}}{\frac{2}{9}\cdot\frac{2}{2}-\frac{1}{18}}=\dfrac{\frac{9-5}{21}}{\frac{4-1}{18}}=\dfrac{\frac{4}{21}}{\frac{3}{18}}=\dfrac{4}{21}\cdot\dfrac{6}{1}=\dfrac{8}{7}$

43. $\dfrac{\frac{25xy^2}{49}}{\frac{15x^2y}{14}}=\dfrac{25xy^2}{49}\cdot\dfrac{14}{15x^2y}=\dfrac{10y}{21x}$

45. $\dfrac{\frac{1}{x}-\frac{1}{y}}{\frac{1}{x}+\frac{1}{y}}\cdot\dfrac{xy}{xy}=\dfrac{y-x}{y+x}$

Cumulative Review Problems

47. $(-2)(3)(-1)(-5)=(-6)(5)=-30$

49. $-25\div 5=-5$

5.6 Exercises

1. $\dfrac{1}{2}+3\cdot\dfrac{1}{4}=\dfrac{2}{4}+\dfrac{3}{4}=\dfrac{5}{4}=1\dfrac{1}{4}$ hr

3. $d=rt\Rightarrow r=\dfrac{d}{t}=\dfrac{310}{5\frac{1}{3}}=310\cdot\dfrac{3}{16}=\dfrac{465}{8}$

$r=58\dfrac{1}{8}$ mph

5. (a) $15\dfrac{2}{3}\cdot\dfrac{2}{2}=15\dfrac{4}{6}$

$\underline{-9\dfrac{1}{2}}\cdot\dfrac{3}{3}=\underline{-9\dfrac{3}{6}}$

$6\dfrac{1}{6}$ gal

(b) $\dfrac{24\text{ mi}}{\text{gal}}\cdot 6\dfrac{1}{6}\text{ gal}=24\cdot\dfrac{37}{6}=148\text{ mi}$

7. $6\cdot 1\dfrac{1}{4}=6\cdot\dfrac{5}{4}=\dfrac{15}{2}=7\dfrac{1}{2}$ cups water

$6\cdot\dfrac{1}{8}=\dfrac{6}{8}=\dfrac{3}{4}$ tsp salt

$6\cdot\dfrac{1}{4}=\dfrac{6}{4}=\dfrac{3}{2}=1\dfrac{1}{2}$ cups cereal

9. $5\cdot 2\dfrac{3}{4}=5\cdot\dfrac{11}{4}=\dfrac{55}{4}=13\dfrac{3}{4}$ yd

11. (a) $\dfrac{240{,}000\text{ mi}}{4\frac{1}{2}\text{ day}}=53{,}333\dfrac{1}{3}\text{ mi/day}$

(b) $\dfrac{240{,}000\text{ mi}}{4\frac{1}{2}\text{ day}}\cdot\dfrac{\text{day}}{24\text{ hr}}=2222\dfrac{2}{9}\text{ mi/hr}$

13. (a) $40-2\cdot 4\dfrac{3}{4}=30\dfrac{1}{2},\ 25-2\cdot 4\dfrac{3}{4}=15\dfrac{1}{2}$

dimensions: $30\dfrac{1}{2}$ ft by $15\dfrac{1}{2}$ ft

(b) $\left(2\cdot 30\dfrac{1}{2}+2\cdot 15\dfrac{1}{2}\right)\cdot 2\dfrac{1}{4}=\$207$

15. $3\cdot 4\cdot 3\dfrac{3}{4}=45$ ft, $\dfrac{45}{8}=5\dfrac{5}{8}\rightarrow 6$ boards

17. $7\dfrac{10}{12}-7\dfrac{8}{12}=\dfrac{2}{12}\rightarrow 2$ in.

19. $7 \text{ ft } 8 \text{ in.} - 6 \text{ ft } 5\frac{5}{8} \text{ in.} = 1 \text{ ft } (8 - 5\frac{5}{8}) \text{ in.}$

$= 1 \text{ ft } 2\frac{3}{8} \text{ in.}$

21. $60 - \frac{1}{4} \cdot 60 - \frac{1}{3} \cdot 60 = 25 \text{ ft}$

Cumulative Review Problems

23. $3x = 12 \Rightarrow x = \frac{12}{3} = 4$

25. $\frac{x}{7} = 6 \Rightarrow x = 7 \cdot 6 = 42$

27. $2(20) + 2(16) = 72 \text{ ft}$

5.7 Exercises

1. When you multiply a nonzero fraction by its reciprocal, the product is 1 .

3. To solve $\frac{x}{6} = 2$, we multiply by 6 on both sides of the equation.

5. $\frac{5x}{5} = x$

7. $\frac{-2x}{-2} = x$

9. $\frac{6 \cdot x}{6} = x$

11. $\frac{-5x}{-5} = x$

13. $\frac{y}{15} = 12 \Rightarrow y = 15(12) = 180$

13. check: $\frac{180}{15} \stackrel{?}{=} 12,\ 12 = 12$

15. $\frac{x}{7} = 31 \Rightarrow x = 7(31) = 217$

check: $\frac{217}{7} \stackrel{?}{=} 31,\ 31 = 31$

17. $\frac{m}{13} = -30 \Rightarrow m = 13(-30) = -390$

check: $\frac{-390}{13} \stackrel{?}{=} -30,\ -30 = -30$

19. $\frac{x}{14} = -6 \Rightarrow x = -6(14) = -84$

check: $\frac{-84}{14} \stackrel{?}{=} -6,\ -6 = -6$

21. $3 = \frac{y}{11} \Rightarrow y = 3(11) = 33$

check: $3 \stackrel{?}{=} \frac{33}{11},\ 3 = 3$

23. $15 = \frac{a}{4} \Rightarrow a = 4(15) = 60$

check: $15 \stackrel{?}{=} \frac{60}{4},\ 15 = 15$

25. $-1 = \frac{m}{30} \Rightarrow m = -1(30) = -30$

check: $-1 \stackrel{?}{=} \frac{-30}{30},\ -1 = -1$

27. $-4 = \frac{a}{-20} \Rightarrow 2 = -4(-20) = 80$

check: $-4 \stackrel{?}{=} \frac{80}{-20},\ -4 = -4$

29. (a) $4x = 52 \Rightarrow x = \frac{52}{4} = 13$

(b) $\frac{x}{4} = 52 \Rightarrow x = 4(52) = 208$

31. (a) $4x = -52 \Rightarrow x = \frac{-52}{4} = -13$

31. (b) $\frac{x}{4}=-52 \Rightarrow x=4(-52)=-208$

33. $\frac{x}{3^2}=2+6\div 3 \Rightarrow \frac{x}{9}=2+2=4$
$x=9(4)=36$

35. $\frac{x}{-7}=-2+3=1 \Rightarrow x=-7\cdot 1=-7$

37. $\frac{y}{2^3}=2\cdot 3+1 \Rightarrow \frac{y}{8}=6+1=7$
$y=8(7)=56$

39. $\frac{3}{4}y=9 \Rightarrow y=\frac{4}{3}\cdot 9=12$

check: $\frac{3}{4}\cdot 12 \overset{?}{=} 9,\ 9=9$

41. $\frac{2}{7}x=12 \Rightarrow x=\frac{7}{2}\cdot 12=42$

check: $\frac{2}{7}\cdot 42 \overset{?}{=} 12,\ 12=12$

43. $\frac{1}{2}x=-15 \Rightarrow x=2(-15)=-30$

check: $\frac{1}{2}(-30) \overset{?}{=} -15,\ -15=-15$

45. $\frac{-5}{8}x=30 \Rightarrow x=\frac{8}{-5}\cdot 30=-48$

check: $\frac{-5}{8}\cdot(-48) \overset{?}{=} 30,\ 30=30$

47. $\frac{-3}{4}y=-12 \Rightarrow y=\frac{4}{-3}\cdot(-12)=16$

check: $\frac{-3}{4}\cdot 16 \overset{?}{=} -12,\ -12=-12$

49. (a) $x-7=21 \Rightarrow x=21+7=28$

(b) $x+12=33 \Rightarrow x=33-12=21$

(c) $5x=3 \Rightarrow x=\frac{3}{5}$

(d) $\frac{x}{5}=11 \Rightarrow x=5(11)=55$

51. (a) $x-7=12 \Rightarrow x=12+7=19$

(b) $3x=2 \Rightarrow x=\frac{2}{3}$

(c) $\frac{x}{6}=9 \Rightarrow x=6(9)=54$

(d) $x+11=34 \Rightarrow x=34-11=23$

Cumulative Review Problems

53. $145\div 11=13$ R2 or $13\frac{2}{11}$

55. $29{,}441=29{,}400$ to nearest hundred

57. $A=LW=\left(4\frac{2}{3}\right)\left(3\frac{3}{8}\right)=\left(\frac{14}{3}\right)\left(\frac{27}{8}\right)$

$A=15\frac{3}{4}\text{ ft}^2$

Chapter 5 Review

1. The reciprocal of $\frac{2}{-9}$ is $\frac{-9}{2}$

2. The reciprocal of 7 is $\frac{1}{7}$

3. The reciprocal of $\frac{-a}{b}$ is $\frac{b}{-a}$

4. $\frac{5}{21}\cdot\frac{3}{15}=\frac{15}{21\cdot 15}=\frac{1}{21}$

5. $\dfrac{-\cancel{6}}{\cancel{35}_{5}} \cdot \dfrac{\cancel{14}^{2}}{\cancel{18}_{3}} = -\dfrac{2}{15}$

6. $\dfrac{\cancel{9}\cancel{x}}{\cancel{15}_{5}} \cdot \dfrac{\cancel{21}^{7}}{\cancel{18}_{2}\,x^{\cancel{2}}} = \dfrac{7}{10x}$

7. $\dfrac{\cancel{8}^{4}x^{\cancel{2}}}{\cancel{25}_{5}} \cdot \dfrac{-\cancel{45}^{\cancel{9}}}{\cancel{18}_{\cancel{9}}\,\cancel{x}} = -\dfrac{4x}{5}$

8. $\dfrac{-5x^4}{\cancel{6}} \cdot \cancel{12}^{2}x^3 = -10x^{4+3} = -10x^7$

9. $\dfrac{9}{14} \div \dfrac{45}{12} = \dfrac{\cancel{9}}{\cancel{14}_{7}} \cdot \dfrac{\cancel{12}^{6}}{\cancel{45}_{5}} = \dfrac{6}{35}$

10. $\dfrac{7}{15} \div \dfrac{-35}{20} = -\dfrac{\cancel{7}}{\cancel{15}_{3}} \cdot \dfrac{\cancel{20}^{4}}{\cancel{35}_{5}} = -\dfrac{4}{15}$

11. $\dfrac{8}{42} \div \dfrac{-22}{7} = -\dfrac{\cancel{8}^{\cancel{4}^{2}}}{\cancel{42}_{\cancel{6}_{3}}} \cdot \dfrac{\cancel{7}}{\cancel{22}_{11}} = -\dfrac{2}{33}$

12. $\dfrac{11x^5}{25} \div \dfrac{3}{5x^2} = \dfrac{11x^5}{\cancel{25}_{5}} \cdot \dfrac{\cancel{5}x^2}{3} = \dfrac{11x^7}{15}$

13. $\dfrac{16x^2}{9} \div \dfrac{24}{6x^4} = \dfrac{\cancel{16}^{4}x^2}{9} \cdot \dfrac{\cancel{6}x^4}{\cancel{24}_{\cancel{4}}} = \dfrac{4x^6}{9}$

14. $A = \dfrac{1}{2}bh = \dfrac{1}{2} \cdot 18(7) = 63 \text{ m}^2$

15. $A = \dfrac{1}{2}bh = \dfrac{1}{2} \cdot 13(20) = 130 \text{ in.}^2$

16. $\dfrac{2}{7} \cdot 3500 = \$1000$

17. $\dfrac{1}{3} \cdot \dfrac{1}{2} = \dfrac{1}{6}$ lb in each container

18. $7 = 7,\ 14 = 2 \cdot 7,\ \text{LCM} = 2 \cdot 7 = 14$

19. $10 = 2 \cdot 5,\ 20 = 2^2 \cdot 5,\ \text{LCM} = 2^2 \cdot 5 = 20$

20. $18 = 2 \cdot 3^2,\ 20 = 2^2 \cdot 5$
$\text{LCM} = 2^2 \cdot 3^2 \cdot 5 = 180$

21. $42 = 2 \cdot 3 \cdot 7,\ 12 = 2^2 \cdot 3$
$\text{LCM} = 2^2 \cdot 3 \cdot 7 = 84$

22. $4x = 2^2 x,\ 8 = 2^3,\ 16x = 2^4 x$
$\text{LCM} = 2^4 x = 16x$

23. $7x = 7x,\ 14x = 2 \cdot 7x,\ 20 = 2^2 \cdot 5$
$\text{LCM} = 2^2 \cdot 5 \cdot 7x = 140x$

24. $18x = 2 \cdot 3^2 x,\ 45x^2 = 3^2 \cdot 5x^2$
$\text{LCM} = 2 \cdot 3^2 \cdot 5x^2 = 90x^2$

25. $20x = 2^2 \cdot 5x,\ 25x^2 = 5^2 x^2$
$\text{LCM} = 2^2 \cdot 5^2 x^2 = 100x^2$

26. $\dfrac{6}{17} - \dfrac{3}{17} = \dfrac{6-3}{17} = \dfrac{3}{17}$

27. $\dfrac{-23}{27} + \dfrac{-11}{27} = \dfrac{-23+(-11)}{27} = -\dfrac{34}{27}$

28. $\dfrac{7}{x} - \dfrac{5}{x} = \dfrac{7-5}{x} = \dfrac{2}{x}$

29. $\frac{x}{7}-\frac{5}{7}=\frac{x-5}{7}$

30. $\frac{5}{6}\cdot\frac{3}{3}+\frac{4}{9}\cdot\frac{2}{2}=\frac{15+8}{18}=\frac{23}{18}$

31. $\frac{15}{32}\cdot\frac{7}{7}-\frac{7}{28}\cdot\frac{8}{8}=\frac{105-56}{224}$
$=\frac{49}{224}$
$=\frac{7}{32}$

32. $\frac{-3}{14}\cdot\frac{3}{3}+\frac{7}{21}\cdot\frac{2}{2}=\frac{-9+14}{42}=\frac{5}{42}$

33. $\frac{5}{2x}\cdot\frac{3}{3}+\frac{8}{3x}\cdot\frac{2}{3}=\frac{15+16}{6x}=\frac{31}{6x}$

34. $\frac{4x}{15}\cdot\frac{3}{3}+\frac{3x}{45}=\frac{12x+3x}{45}=\frac{15x}{45}=\frac{x}{3}$

35. $\frac{3x}{14}\cdot\frac{3}{3}-\frac{5x}{42}=\frac{9x-5x}{42}=\frac{4x}{42}=\frac{2x}{21}$

36. $10\frac{1}{2}=10\frac{5}{10}$
$+3\frac{4}{5}=+3\frac{8}{10}$
$13\frac{13}{10}=14\frac{3}{10}$

37. $12\frac{7}{9}\quad 12\frac{7}{9}$
$+6\frac{2}{3}=+6\frac{6}{9}$
$18\frac{13}{9}=19\frac{4}{9}$

38. $11\frac{1}{5}=11\frac{5}{25}=10\frac{30}{25}$
$-6\frac{11}{25}\quad -6\frac{11}{25}$
$4\frac{19}{25}$

39. $25=24\frac{14}{14}$
$-16\frac{5}{14}$
$8\frac{9}{14}$

40. $4\frac{1}{2}\cdot 2\frac{2}{9}=\frac{\cancel{9}}{\cancel{2}}\cdot\frac{\overset{10}{\cancel{20}}}{\cancel{9}}=10$

41. $2\frac{3}{4}\div 1\frac{3}{7}=\frac{11}{4}\cdot\frac{7}{10}=\frac{77}{40}$

42. $4\frac{2}{5}\div 8\frac{1}{3}=\frac{22}{5}\cdot\frac{3}{25}=\frac{66}{125}$

43. $-12\div\frac{2}{3}=-12\cdot\frac{3}{2}=-18$

44. $75-2\cdot 20\frac{1}{2}-2\cdot 10\frac{1}{4}=13\frac{1}{2}$ ft

45. $\frac{1}{4}\cdot 7\frac{1}{2}=\frac{1}{4}\cdot\frac{15}{2}=\frac{15}{8}$
$=1\frac{7}{8}$ ft

46. $\frac{3}{4}+\frac{1}{2}\cdot\frac{2}{7}=\frac{3}{4}\cdot\frac{7}{7}+\frac{1}{7}\cdot\frac{4}{4}$
$=\frac{21+4}{28}=\frac{25}{28}$

47. $\left(\frac{3}{4}\right)^2 + \frac{1}{8} \div \frac{1}{2} = \frac{9}{16} + \frac{1}{8} \cdot \frac{2}{1}$
$= \frac{9}{16} + \frac{1}{4} \cdot \frac{4}{4} = \frac{19+4}{16} = \frac{13}{16}$

48. $\dfrac{\frac{2}{3}}{\frac{1}{9}} = \frac{2}{3} \cdot \frac{9}{1} = 6$

49. $\dfrac{\frac{x^2}{2}}{\frac{x}{4}} = \frac{x^{\not 2}}{\not 2} \cdot \frac{\overset{2}{\not 4}}{\not x} = 2x$

50. $\dfrac{\frac{x}{12}}{\frac{x^2}{20}} = \frac{\not x}{\underset{3}{\not 12}} \cdot \frac{\overset{5}{\not 20}}{x^{\not 2}} = \frac{5}{3x}$

51. $\dfrac{\frac{1}{2} \cdot \frac{2}{2} + \frac{1}{4}}{\frac{2}{3} \cdot \frac{3}{3} - \frac{1}{9}} = \dfrac{\frac{3}{4}}{\frac{5}{9}} = \frac{3}{4} \cdot \frac{9}{5} = \frac{27}{20}$

52. $\dfrac{\frac{1}{3} + \frac{2}{5}}{\frac{1}{5} + \frac{1}{10}} \cdot \frac{30}{30} = \frac{10+12}{6+3} = \frac{22}{9}$

53. (a) $38\frac{1}{4} - 2 \cdot 4\frac{1}{2} = 29\frac{1}{4}$

$25 - 2 \cdot 4\frac{1}{2} = 16$

dimensions: $29\frac{1}{4}$ ft by 16 ft

(b) $\left(2 \cdot 29\frac{1}{4} + 2 \cdot 16\right) \cdot 3\frac{1}{4} = \$294\frac{1}{8}$

54. $2\frac{1}{2} - \frac{7}{8} = \frac{5}{2} \cdot \frac{4}{4} - \frac{7}{8}$
$= \frac{20-7}{8} = \frac{13}{8} = 1\frac{5}{8}$ in.

55. $\frac{x}{3} = 10 \Rightarrow x = 3(10) = 30$

56. $\frac{y}{2} = -5 \Rightarrow y = -5(2) = -10$

57. $\frac{x}{-6} = -2 \Rightarrow x = (-6)(-2) = 12$

58. $\frac{2}{3}y = 16 \Rightarrow y = \frac{3}{2} \cdot 16 = 24$

59. $\frac{-3}{4}x = 18 \Rightarrow x = \frac{4}{-3} \cdot 18 = -24$

60. $\frac{1}{3}y = 4 \Rightarrow y = 3(4) = 12$

61. $\frac{x}{2^3} = 2 \cdot 6 + 1 \Rightarrow \frac{x}{8} = 12 + 1 = 13$
$x = 8(13) = 104$

62. $\frac{y}{3^2} = 1 + 8 \div 2 \Rightarrow \frac{y}{9} = 1 + 4 = 5$
$y = 9(5) = 45$

Chapter 5 Test

1. $\frac{\not 2}{\not 3} \cdot \frac{\overset{3}{\not 9}}{\underset{5}{\not 10}} = \frac{3}{5}$

2. $\frac{-1}{4} \cdot \frac{2}{5} = -\frac{2}{20} = -\frac{1}{10}$

3. $\frac{\not{7}x^2}{\not{8}}\cdot\frac{\not{16}^{2}x}{\not{14}_{2}}=x^3$

4. $\frac{-1}{2}\div\frac{1}{4}=-\frac{1}{\not{2}}\cdot\frac{\not{4}^{2}}{1}=-2$

5. $\frac{2x^8}{5}\div\frac{22x^4}{15}=\frac{2x^8}{5}\cdot\frac{15}{22x^4}=\frac{3x^4}{11}$

6. Each 10 ft board will make 2 shelves with 3 ft $=10-2\cdot3\frac{1}{2}$ left over. Therefore she needs 5 boards to make 10 shelves.

7. 9, 18, 27, 36, 45

8. $14=2\cdot7,\ 21=3\cdot7$
$\text{LCM}=2\cdot3\cdot7=42$

9. $5a=5a,\ 10a^4=2\cdot5a^4,\ 20a^2=2^2\cdot5a^2$
$\text{LCM}=2^2\cdot5a^4=20a^4$

10. $5=5,\ 7=7,\ 51=3\cdot17$
$\text{LCM}=3\cdot5\cdot7\cdot17=1785$

11. $30=2\cdot3\cdot5,\ 4=2^2$
$\text{LCD}=2^2\cdot3\cdot5=60$

12. $\frac{12}{11}+\frac{3}{11}=\frac{12+3}{11}=\frac{15}{11}=1\frac{4}{11}$

13. $\frac{1}{5a}+\frac{3}{a}\cdot\frac{5}{5}=\frac{1+15}{5a}=\frac{16}{5a}$

14.
$$\begin{array}{rr} 4\frac{5}{6} & 4\frac{5}{6} \\ +3\frac{1}{3}\cdot\frac{2}{2}= & +3\frac{2}{6} \\ \hline & 7\frac{7}{6}=8\frac{1}{6} \end{array}$$

15. $\frac{2}{3}\cdot\frac{5}{5}+\frac{3}{5}\cdot\frac{3}{3}=\frac{10}{15}+\frac{9}{15}=\frac{19}{15}=1\frac{4}{15}$

16. $\frac{8}{12}-\frac{2}{3}\cdot\frac{4}{4}=\frac{8-8}{12}=\frac{0}{12}=0$

17.
$$\begin{array}{rrr} 10\frac{1}{8}\cdot\frac{3}{3}= & 10\frac{3}{24}= & 9\frac{27}{24} \\ -2\frac{2}{3}\cdot\frac{8}{8}= & -2\frac{16}{24} & -2\frac{16}{24} \\ \hline & & 7\frac{11}{24} \end{array}$$

18. $\frac{10x}{15}-\frac{x}{5}\cdot\frac{3}{3}=\frac{10x-3x}{15}=\frac{7x}{15}$

19. $(-4)\cdot5\frac{1}{3}=(-4)\cdot\frac{16}{3}=-\frac{64}{3}=-21\frac{1}{3}$

20. $\frac{1}{4}\cdot3\frac{1}{2}=\frac{1}{4}\cdot\frac{7}{2}=\frac{7}{8}$

21. $1\frac{2}{3}\div3=\frac{5}{3}\cdot\frac{1}{3}=\frac{5}{9}$

22. $2\frac{1}{2}\div\left(-\frac{3}{5}\right)=-\frac{5}{2}\cdot\frac{5}{3}=-\frac{25}{6}=-4\frac{1}{6}$

23. $\frac{2}{3}+\frac{1}{2}\cdot\frac{1}{4}=\frac{2}{3}\cdot\frac{8}{8}+\frac{1}{8}\cdot\frac{3}{3}=\frac{16+3}{24}=\frac{19}{24}$

24. $\dfrac{\frac{x}{2}}{\frac{x}{4}} = \dfrac{\cancel{x}}{\cancel{2}} \cdot \dfrac{\overset{2}{\cancel{4}}}{\cancel{x}} = 2$

25. $\dfrac{\frac{1}{3}+\frac{1}{2}}{\frac{5}{9}-\frac{1}{3}} \cdot \dfrac{18}{18} = \dfrac{6+9}{10-6} = \dfrac{15}{4} = 3\dfrac{3}{4}$

26. $\dfrac{x}{-3} = 6 \Rightarrow x = -3(6) = -18$

27. $\dfrac{4}{7}y = 28 \Rightarrow y = \dfrac{7}{4} \cdot 28 = 49$

28. $\dfrac{x}{4} = -20 \Rightarrow x = 4(-20) = -80$

29. (a) $98\dfrac{1}{2}°\text{F} + 3°\text{F} - 1\dfrac{1}{2}°\text{F} = 100°\text{F}$

(b) $100°\text{F} - 98\dfrac{1}{2}°\text{F} = 1\dfrac{1}{2}°\text{F}$

Cumulative Test for Chapters 1-5

1. $3401 = 3000 + 400 + 1$

2. $2 + x + 8 = 10 + x$

3. $5x - 3x + x + 5 = (5 - 3 + 1)x + 5 = 3x + 5$

4. $-8r + 3 - 5r - 8 = (-8 - 5)r - 5 = -13r - 5$

5. $4 + x = 15 \Rightarrow x = 15 - 4 = 11$

6. $3x = 15 \Rightarrow x = \dfrac{15}{3} = 5$

7. $x - 8 = 18 \Rightarrow x = 18 + 8 = 26$

8. $(3x^2)(x^3)(x) = 3x^{2+3+1} = 3x^6$

9. $5(x+3) = 5x + 5(3) = 5x + 15$

10. $8^2 - 10 + 4 = 64 - 10 + 4 = 54 + 4 = 58$

11. $-4 + 6 - 9 = 2 - 9 = -7$

12. $(-15) \div 3 = -5$

13. $(-7)(-1) = 7$

14. $(-10x^2)(5x) = -50x^{2+1} = -50x^3$

15. $$\begin{aligned} 3 - 12 \div (-2) + 4^2 &= 3 - 12 \div (-2) + 16 \\ &= 3 - (-6) + 16 \\ &= 9 + 16 \\ &= 25 \end{aligned}$$

16. (a) $72 = 2 \cdot 2 \cdot 2 \cdot 3 \cdot 3$ composite
(b) 19 prime

17. $4x = 2^2x,\ 8x^3 = 2^3x^3,\ 12xy = 2^2 \cdot 3xy$
LCM $= 2^3 \cdot 3x^3y = 24x^3y$

18. $24 = 2^3 \cdot 3,\ 30 = 2 \cdot 3 \cdot 5$
LCM $= 2^3 \cdot 3 \cdot 5 = 120$

19. $\dfrac{\frac{2}{x^2}}{\frac{8}{x}} \cdot \dfrac{x^2}{x^2} = \dfrac{2}{8x} = \dfrac{1}{4x}$

20. $\dfrac{\frac{3}{4}-\frac{1}{2}}{\frac{1}{6}+\frac{2}{3}} \cdot \dfrac{12}{12} = \dfrac{9-6}{2+8} = \dfrac{3}{10}$

21. $11\frac{1}{5}\cdot\frac{3}{3}= \quad 11\frac{3}{15}=10\frac{18}{15}$

$\underline{-3\frac{2}{3}}\cdot\frac{5}{5}= \qquad \underline{-3\frac{10}{15}}$

$7\frac{8}{15}$

22. $\frac{1}{6x}+\frac{3}{x}\cdot\frac{6}{6}=\frac{1+18}{6x}=\frac{19}{6x}$

23. $\frac{5x}{2}\cdot\frac{5}{5}-\frac{x}{10}=\frac{25x-x}{10}=\frac{24x}{10}=\frac{12x}{5}$

24. $\frac{\cancel{2}x^4}{\cancel{5}x}\cdot\frac{\overset{2}{\cancel{10}}x^2}{\underset{2}{\cancel{4}}}=x^{4+2-1}=x^5$

25. $-\frac{1}{6}\div\left(-\frac{2}{3}\right)=\frac{1}{6}\cdot\frac{3}{2}=\frac{1}{4}$

26. $7\frac{1}{2}\div\left(-\frac{3}{8}\right)=-\frac{15}{2}\cdot\frac{8}{3}=-20$

27. $\frac{1}{2}\cdot\frac{5}{5}+\frac{3}{5}\cdot 2\cdot\frac{2}{2}=\frac{5+12}{10}=\frac{17}{10}=1\frac{7}{10}$

28. $\frac{-2}{9}x=12\Rightarrow x=\frac{9}{-2}\cdot 12=-54$

29. $\frac{x}{-6}=5\Rightarrow x=(-6)(5)=-30$

30. $12\frac{1}{2}-9\frac{1}{3}=12\frac{3}{6}-9\frac{2}{6}=3\frac{1}{6}$ gal

31. $\frac{2}{3}\cdot 5\frac{1}{4}=\frac{2}{3}\cdot\frac{21}{4}=\frac{21}{6}=3\frac{1}{2}$ lb

Chapter 6

Pretest Chapter 6

1. The terms of $ab^3 - 2a + 3b - 12$ are
$ab^3, \ -2a, \ 3b, \ -12$.

2. $-(-3x + 2y - 8) = 3x - 2y + 8$

3. $(5x^2 + 3x - 1) + (7x^2 - 2x + 3)$
$= 5x^2 + 3x - 1 + 7x^2 - 2x + 3$
$= 12x^2 + x + 2$

4. $(8y + 4) - (-3y + 1) = 8y + 4 + 3y - 1$
$= 11y + 3$

5. $6x - (-3x + 4) + 2(3x^2 + 5x - 1)$
$= 6x + 3x - 4 + 6x^2 + 10x - 2$
$= 6x^2 + 19x - 6$

6. $(-3a)(4a + 6) = (-3a)(4a) + (-3a)(6)$
$= -12a^2 - 18a$

7. $(-3x^3)(4x^2 + 3x - 5)$
$= (-3x^3)(4x^2) + (-3x^3)(3x) - (-3x^3)(5)$
$= -12x^5 - 9x^4 + 15x^3$

8. $(x + 2)(x - 7) = x^2 - 7x + 2x - 14$
$= x^2 - 5x - 14$

9. $(x + 1)(2x^2 + 3x - 4)$
$= 2x^3 + 3x^2 - 4x + 2x^2 + 3x - 4$
$= 2x^3 + 5x^2 - x - 4$

10. Height of tree $= t$
Height of building $= 2t$

11. length $= x$, width $= 3x - 6$

12. (a) Second quarter sales $= s$
first quarter sales $= 4s$
third quarter sales $= 3s + 400$
(b) $4s - (3s + 400)$
(c) $4s - (3s + 400) = 4s - 3s - 400$
$= s - 400$

13. $9x + 3 = 3(3x + 1)$

14. $10x^2y - 15xy = 5xy(2x - 3)$

15. $4a - 12b = 4(a - 3b)$

16. $7x^3y - 21x^2y^2 + 14xy = 7xy(x^2 - 3xy + 2)$

6.1 Exercises

1. To subtract two polynomials we add the opposite of the <u>second</u> polynomial to the <u>first</u> polynomial.

3. Because we can replace this negative sign with a -1 and then multiply each term by -1. Multiplying a term by -1 has the effect of changing the sign.

5. The terms of $2z^2 + 4z - 2y^4 + 3$ are
$2z^2, \ 4z, \ -2y^4, \ 3$

7. The terms of $6x^6 - 3x^3 - 3y - 7$ are
$6x^6, \ -3x^3, \ -3y, \ -7$

9. $(7y - 3) + (-4y + 6) = 7y - 3 - 4y + 6$
$= 3y + 3$

11. $(2a^2 + (-3a) + 5) + (4a - 3)$
$= 2a^2 - 3a + 5 + 4a - 3 = 2a^2 + a + 2$

13. $(5y^2+2y-5)+(-4y^2+(-8y)+2)$
$=5y^2+2y-5-4y^2-8y+2$
$=y^2-6y-3$

15. $-(5x+2y)=-5x-2y$

17. $-(-8x+9)=8x-9$

19. $-(-3x+6z-5y)=3x-6z+5y$

21. $(7x+3)-(-4x-6)=7x+3+4x+6$
$=11x+9$

23. $(-8a+5)-(4a-3)=-8a+5-4a+3$
$=-12a+8$

25. $(3y^2+4y-6)-(4y^2-6y-9)$
$=3y^2+4y-6-4y^2+6y+9$
$=-y^2+10y+3$

27. $(2x^2+6x-5)-(6x^2-4x-8)$
$=2x^2+6x-5-6x^2+4x+8$
$=-4x^2+10x+3$

29. $(-6z^2+9z-1)-(3z^2+8z-7)$
$=-6z^2+9z-1-3z^2-8z+7$
$=-9z^2+z+6$

31. $(2a^2+9a-1)-(-5a^2-8a-4)$
$=2a^2+9a-1+5a^2+8a+4$
$=7a^2+17a+3$

33. $(-6x^2-6x-1)-(3x^2+8x+4)$
$=-6x^2-6x-1-3x^2-8x-4$
$=-9x^2-14x-5$

35. $(x^2-x-1)+(-x^2+4x+6)-(x^2+1)$
$=x^2-x-1-x^2+4x+6-x^2-1$
$=-x^2+3x+4$

37. $(-4x^2-1)-(3x^2+8x+2)+(-x^2-9)$
$=-4x^2-1-3x^2-8x-2-x^2-9$
$=-8x^2-8x-12$

39. $(6x+9)-(-4x^2+7x+1)-(x^2-7)$
$=6x+9+4x^2-7x-1-x^2+7$
$=3x^2-x+15$

41. $(4x^2-5x-8)-(3x^2-8x-1)+(x^2-9)$
$=4x^2-5x-8-3x^2+8x+1+x^2-9$
$=2x^2+3x-16$

43. $(4x^2+6x-2)-(3x^2+7x+1)+(x^2-1)$
$=4x^2+6x-2-3x^2-7x-1+x^2-1$
$=2x^2-x-4$

45. $(-7x^2+3x-4)-(5x^2+7x-1)+(x^2-8)$
$=-7x^2+3x-4-5x^2-7x+1+x^2-8$
$=-11x^2-4x-11$

47. $5x-2(4x^2+3x-2)-(2x^2+8x-1)$
$=5x-8x^2-6x+4-2x^2-8x+1$
$=-10x^2-9x+5$

49. $4(-9x-6)-(3x^2-7x+1)+4x$
$=-36x-24-3x^2+7x-1+4x$
$=-3x^2-25x-25$

51. $(ax+3)+(2x^2+5x-6)+(8x-2)$
$=2x^2-10x-5$
$ax+3+2x^2+5x-6+8x-2$
$=2x^2-10x-5$
$2x^2+(a+5+8)x-5-2x^2-10x-5$
$\Rightarrow a+13=-10 \Rightarrow a=-23$

Cumulative Review Problems

53. $(2x)(4x)=8x^2$

55. $(-4x)(2x^2)=-8x^3$

57. $3\frac{1}{2}\cdot\frac{7}{7}-2\frac{2}{7}\cdot\frac{2}{2}=3\frac{7}{14}-2\frac{4}{14}=1\frac{3}{14}$ mi

6.2 Exercises

1. The sign of the last term, +4, is wrong. It should be −4.

3. (a) A <u>monomial</u> is a polynomial with 1 term.
(b) A <u>binomial</u> is a polynomial with 2 terms.
(c) A <u>trinomial</u> is a polynomial with 3 terms.

5. $8(2x^2+3x-5)=16x^2+24x-40$

7. $-5(2y^2-3y+6)=-10y^2+15y-30$

9. $-9a(a+1)=-9a^2-9a$

11. $-7x(3x-2y+4)=-21x^2+14xy-28x$

13. $-3a(2a-6b-5)=-6a^2+18ab+15a$

15. $-4x(-3x+5y-4)=12x^2-20xy+16x$

17. $-7x^3(x-3)=-7x^4+21x^3$

19. $-6x^5(x^2+4)=-6x^7-24x^5$

21. $(x^3+2)(-5x^2)=-5x^5-10x^2$

23. $(3y+2)(4y^2+3y-2)$
$=12y^3+9y^2-6y+8y^2+6y-4$
$=12y^3+17y^2-4$

25. $(x-2)(x^2+3x+1)$
$=x^3+3x^2+x-2x^2-6x-2$
$=x^3+x^2-5x-2$

27. $(x-1)(2x^2-3x-2)$
$=2x^3-3x^2-2x-2x^2+3x+2$
$=2x^3-5x^2+x+2$

29. $(x+6)(x+7)=x^2+7x+6x+42$
$=x^2+13x+42$

31. $(x+3)(x+9)=x^2+9x+3x+27$
$=x^2+12x+27$

33. $(a+6)(a+2)=a^2+2a+6a+12$
$=a^2+8a+12$

35. $(y+4)(y-8)=y^2-8y+4y-32$
$=y^2-4y-32$

37. $(x+2)(x-4)=x^2-4x+2x-8$
$=x^2-2x-8$

39. $(z+5)(z-2)=z^2-2z+5z-10$
$=z^2+3z-10$

41. $(x-2)(x+4) = x^2 + 4x - 2x - 8$
$= x^2 + 2x - 8$

43. $(y-2)(y+7) = y^2 + 7y - 2y - 14$
$= y^2 + 5y - 14$

45. $(2x+1)(x+2) = 2x^2 + 4x + x + 2$
$= 2x^2 + 5x + 2$

47. $(3x-3)(x-1) = 3x^2 - 3x - 3x + 3$
$= 3x^2 - 6x + 3$

49. $(2y-1)(y+2) = 2y^2 + 4y - y - 2$
$= 2y^2 + 3y - 2$

51. $(2y+1)(y-2) = 2y^2 - 4y + y - 2$
$= 2y^2 - 3y - 2$

53. (a) $(z+5)(z+1) = z^2 + z + 5z + 5$
$= z^2 + 6z + 5$
(b) $(z-5)(z+1) = z^2 + z - 5z - 5$
$= z^2 - 4z - 5$
(c) $(z-5)(z-1) = z^2 - z - 5z + 5$
$= z^2 - 6z + 5$
(d) $(z+5)(z-1) = z^2 - z + 5z - 5$
$= z^2 + 4z - 5$

55. $a(2x-3) = -14x + 21$
$2ax - 3a = -14x + 21 \Rightarrow 2a = -14$
$a = -7$

Cumulative Review Problems

57. $-2-(-5)+(-7) = 3+(-7) = -4$

59. $-3x^2 + 5x^2 + (-6x^2) = 2x^2 + (-6x^2)$
$= -4x^2$

61. $\frac{3}{4} \cdot 1200 = 900$ people

6.3 Exercises

1. $x =$ Mark's age and $3x =$ Jesse's.

3. $x =$ Leslie's running record,
$x - 20 =$ Alice's running record, and
$x - 50 =$ Shannon's running record.

5. The bother's salary $= B$,
Victor's salary $= B - 95$

7. Profit the second quarter $= S$,
profit the first quarter $= 31{,}100 + S$

9. width $= x$, length $= 2x$

11. The height of the tree $= x$,
the height of a pole $= \frac{1}{2}x$

13. length $= x$, width $= 2x - 13$

15. Late arrivals this year $= x$,
late arrivals last year $= 2x - 420$

17. width $= w$, length $= 2w$, height $= 3w$

19. $J =$ Janet's height
$J + 2 =$ Wendy's height
$J - 3 =$ Marci's height

21. $F =$ length of first side
$F + 4 =$ length of second side
$2F - 10 =$ length of third side

23. (a) height of tree $= t$,
height of building $= 4t$

23. (b) $4t - t$
(c) $4t - t = 3t$

25. (a) Number of coins in John's collection $= x$, Number of coins in Carina's collection $= x + 10$, Number of coins in Karla's collection $= x - 6$
(b) $(x+10)+x-(x-6)$
(c) $(x+10)+x-(x-6)$
$= x+10+x-x+6 = x+16$

27. (a) Sam's salary $= x$,
Vu's salary $= x + 125$,
Evan's salary $= x - 80$
(b) $(x+125)+x-(x-80)$
(c) $(x+125)+x-(x-80)$
$= x+125+x-x+80 = x+205$

29. (a) width $= x$, height $= x + 8$,
length $= 2x - 2$
(b) $x+(x+8)+(2x-2)$
(c) $x+(x+8)+(2x-2)$
$= x+x+8+2x-2 = 4x+6$

31. We can <u>simplify</u> $3x+1+4x$.

33. We can <u>simplify and solve</u> $4+2x+6x=11$.

35. We can solve $4x-2$. <u>false</u>

Cumulative Review Problems

37. $2x = 44 \Rightarrow x = \dfrac{44}{2} = 22$

39. $\dfrac{m}{7} = -5 \Rightarrow m = 7(-5) = -35$

Putting Your Skills to Work, Problems for Individual Investigation

1. (a)

Salary Level	Monthly Salary	Formula
Level 1		s
Level 2		$s+300$
Level 3		$s+600$
Level 4		$s+900$

(b)

Salary Level	Monthly Salary	Formula
Level 1	\$1600	s
Level 2	\$1900	$s+300$
Level 3	\$2200	$s+600$
Level 4		$s+900$

2. (a)

Salary Level	Salary	Formula
Level 1	\$1500	x
Level 2	1700	$x+200$
Level 3	1900	$x+400$
Level 4		$x+500$
Level 5		$x+800$

(b)

Salary Level	Salary	Formula
Level 1	\$1500	x
Level 2	1700	$x+200$
Level 3	1900	$x+400$
Level 4		$x+500$
Level 5	2300	$x+800$

3. Level 5: $x+800$, $1650+800=2450$

Putting Your Skills to Work, Problems for Group Investigation and Cooperative Study

1. There are 3 ways to set up the salary schedule.

Level	Salary	Formula
Level 1	1600	s
Level 2	2600	$s+1000$

Level	Salary	Formula
Level 1	1600	s
Level 2	1800	$s+200$
Level 3	2000	$s+400$
Level 4	2200	$s+600$
Level 5	2400	$s+800$
Level 6	2600	$s+1000$

Level	Salary	Formula
Level 1	1600	s
Level 2	2100	$s+500$
Level 3	2600	$s+1000$

2. (a) Answers may vary.

	First Arrangement
Front	A 1 A 2
Middle	7 yr 9 yr
Back	12 yr 14 yr empty

	Second Arrangement
Front	A 1 A 2
Middle	9 yr 7 yr
Back	empty 12 yr 14 yr

2. (a)

	Third Arrangement
Front	A 2 14 yr
Middle	A 1 12 yr
Back	7 yr 9 yr empty

(b)

	First Arrangement
Front	A 1 A 2
Middle	7 yr 14 yr
Back	12 yr 9 yr empty

	Second Arrangement
Front	A 1 14 yr
Middle	9 yr 12 yr
Back	empty A 2 7 yr

6.4 Exercises

1. The sign in the binomial should be $-$, not $+$.

3. $2(\boxed{x}+\boxed{2})=2x+4$

5. $5(\boxed{2x}+\boxed{y})=10x+5y$

7. $3x(\boxed{2x}-\boxed{4})=6x^2-12x$

9. $3x(\boxed{6x^2}-\boxed{x}-\boxed{3})=18x^3-3x^2-9x$

11. $2ab(\boxed{4a}-\boxed{2a^2b}-\boxed{1})$
$=8a^2b-4a^3b^2-2ab$

13. (a) $2x-10=2(x\boxed{-}10)$
(b) $2x+10=2(x\boxed{+}10)$

15. $3x^2y^2 - 6xy^2 + 9xy^3 = 3xy^2(x \boxed{-} 2 \boxed{+} 3y)$

17. $9 = 3^2,\ 27 = 3^3,\ \text{GCF} = 3^2 = 9$

19. $8 = 2^3,\ 12 = 2^2 \cdot 3,\ \text{GCF} = 2^2 = 4$

21. $3a - 12 = 3(a - 4)$
check: $3(a - 4) = 3a - 3 \cdot 4 = 3a - 12$

23. $4y + 2 = 2(2y + 1)$
check: $2(2y + 1) = 2(2y) + 2(1) = 4y + 2$

25. $6y + 6 = 6(y + 1)$
check: $6(y + 1) = 6y + 6(1) = 6y + 6$

27. $15a - 20 = 5(3a - 4)$
check: $5(3a - 4) = 15a - 5 \cdot 4 = 15a - 20$

29. $4a + 10b = 2(2a + 5b)$
check: $2(2a + 5b) = 4a + 10b$

31. $3m + 15n = 3(m + 5n)$
check: $3(m + 5n) = 3m + 15n$

33. $7x + 14y + 28 = 7(x + 2y + 4)$
check: $7(x + 2y + 4) = 7x + 14y + 28$

35. $8a + 18b - 6 = 2(4a + 9b - 3)$
check: $2(4a + 9b - 3) = 8a + 18b - 6$

37. (a) a and c
(b) a is 3 and c is 2
(c) a^3c^2

39. $2a^2 - 4a = 2a(a - 2)$

41. $5x^2 + 10x^3 = 5x^2(1 + 2x)$

43. $4ab - b^2 = b(4a - b)$

45. $2x^2y + 4xy = 2xy(x + 2)$

47. $5x + 10xy = 5x(1 + 2y)$

49. $5x - 10xy = 5x(1 - 2y)$

51. $7x^2y^3 + 14xy = 7xy(xy^2 + 2)$

53. $9xy^3 - 3xy = 3xy(3y^2 - 1)$

55. $2x^3y^3 - 8x^2y^2 = 2x^2y^2(xy - 4)$

57. $12a^2b^2 - 6a^2 = 6a^2(2b^2 - 1)$

59. $6x^3 - 3x^2 + 12x = 3x(2x^2 - x + 4)$

61. $4y^4 + 8y^3 - 4y^2 = 4y^2(y^2 + 2y - 1)$

63. $6x^2y + 2xy + 4x = 2x(3xy + y + 2)$

65. (a) $-2(x - 5y) = -2x + 10y$
(b) $2(-x + 5y) = -2x + 10y$
(c) The products are the same.

Cumulative Review Problems

67. $3 = 3,\ 4 = 2^2,\ 2 = 2,\ \text{LCD} = 2^2 \cdot 3 = 12$

69. $2x = 2x,\ x = x,\ \text{LCD} = 2x$

71. $10 - 2 \cdot 2\frac{2}{3} - 3\frac{1}{4}$

$= 10 - 2 \cdot \frac{8}{3} - \frac{13}{4}$

$= 10 \cdot \frac{12}{12} - \frac{16}{3} \cdot \frac{4}{4} - \frac{13}{4} \cdot \frac{3}{3}$

$= \frac{120 - 64 - 39}{12} = \frac{17}{12} = 1\frac{5}{12}$ in.

Chapter 6 Review Problems

1. The terms in $2x^2+5x-3z^3+4$ are
$2x^2,\ +5x,\ -3z^3,\ +4$

2. The terms in $8a^5-7b^3-5b-4$ are
$8a^5,\ -7b^3,\ -5b,\ -4$

3. $-(3a-4)=-3a+4$

4. $-(-3y+2z-3)=3y-2z+3$

5. $(-3x+9)+(5x-2)=-3x+9+5x-2$
$=2x+7$

6. $(4x+8)-(8x-2)=4x+8-8x+2$
$=-4x+10$

7. $(9a^2-3a+5)-(-4a^2-6a-1)$
$=9a^2-3a+5+4a^2+6a+1$
$=13a^2+3a+6$

8. $(-4x^2-3)-(3x^2+7x+1)+(-x^2-4)$
$=-4x^2-3-3x^2-7x-1-x^2-4$
$=-8x^2-7x-8$

9. $2(-2x^2+2)-(3x^2+5x-6)$
$=-4x^2+4-3x^2-5x+6$
$=-7x^2-5x+10$

10. $4x-(x^2+2x)+3(3x^2-6x+4)$
$=4x-x^2-2x+9x^2-18x+12$
$=8x^2-16x+12$

11. $(-4)(6x^2-8x+5)=-24x^2+32x-20$

12. $(-2y)(y-6)=-2y^2+12y$

13. $(3x)(9x-3y+2)=27x^2-9xy+6x$

14. $(-5n)(-4n-9m-7)$
$=20n^2+45mn+35n$

15. $(4x^2)(x^4-4)=4x^6-16x^2$

16. $(x^4)(x^5-2x-3)=x^9-2x^5-3x^4$

17. $(z-4)(5z)=5z^2-20z$

18. $(y+10)(-6y)=-6y^2-60y$

19. $(x^3-6x)(4x^2)=4x^5-24x^3$

20. $(x-2)(2x^2+3x-1)$
$=2x^3+3x^2-x-4x^2-6x+2$
$=2x^3-x^2-7x+2$

21. $(y+5)(3y^2-2y+3)$
$=3y^3-2y^2+3y+15y^2-10y+15$
$=3y^3+13y^2-7y+15$

22. $(y-1)(-3y^2+4y+5)$
$=-3y^3+4y^2+5y+3y^2-4y-5$
$=-3y^3+7y^2+y-5$

23. $(2x+3)(x^2+3x-1)$
$=2x^3+6x^2-2x+3x^2+9x-3$
$=2x^3+9x^2+7x-3$

24. $(x+2)(x+4)=x^2+4x+2x+8$
$=x^2+6x+8$

25. $(y+4)(y-7)=y^2-7y+4y-28$
$=y^2-3y-28$

26. $(x-2)(3x+4) = 3x^2 + 4x - 6x - 8$
$= 3x^2 - 2x - 8$

27. $(x-3)(5x-6) = 5x^2 - 6x - 15x + 18$
$= 5x^2 - 21x + 18$

28. Profit for first quarter $= x$,
profit for third quarter $= 22{,}300 + x$

29. length $= L$, width $= L - 22$

30. Measure of $\angle b = x$,
measure of $\angle a = x + 30$,
measure of $\angle c = 2x$

31. Senior citizen tickets sold $= x$,
adult tickets sold $= 3x$,
children's tickets sold $= x + 40$

32. (a) Erin's salary $= x$,
Phoebe's salary $= x + 145$,
Kelly's salary $= x - 60$
(b) $x + (x + 145) - (x - 60)$
(c) $x + (x + 145) - (x - 60)$
$= x + x + 145 - x + 60 = x + 205$

33. (a) Number of coins in Jessica's collection $= x$,
Number of coins in Natasha's collection $= x + 20$,
Number of coins in Le Mar's collection $= x - 15$
(b) $x + (x + 20) - (x - 15)$
(c) $x + (x + 20) - (x - 15)$
$= x + x + 20 - x + 15 = x + 35$

34. (a) first angle $= x$, second angle $= 2x$,
third angle $= x + 10$
(b) $x + 2x + (x + 10)$

34. (c) $x + 2x + (x + 10) = 3x + x + 10$
$= 4x + 10$

35. (a) width $= x$, height $= x + 7$,
length $= 3x - 4$
(b) $x + (x + 7) + (3x - 4)$
(c) $x + (x + 7) + (3x - 4)$
$= x + x + 7 + 3x - 4 = 5x + 3$

36. $2x - 14 = 2(x - 7)$

37. $5x + 15 = 5(x + 3)$

38. $3a + 9b = 3(a + 3b)$

39. $4y + 8z = 4(y + 2z)$

40. $6xy^2 - 12xy = 6xy(y - 2)$

41. $8a^2b - 16ab = 8ab(a - 2)$

42. $10x^3y^2 + 5x^2y^2 = 5x^2y^2(2x + 1)$

43. $4y^3 - 6y^2 + 2y = 2y(2y^2 - 3y + 1)$

44. $3a^2b^2 - 6ab + 12a^2b = 3ab(ab - 2 + 4a)$

45. $2x^2y^3 + 4xy^2 - 10x^2y^2$
$= 2xy^2(xy + 2 - 5x)$

Chapter 6 Test

1. The terms in $x^2y - 2x^2 + 3y - 5x$ are
$x^2y,\ -2x^2,\ +3y,\ -5x$

2. $-(4x - 2y - 6) = -4x + 2y + 6$

3. $(-5x + 3) + (-2x + 4) = -5x + 3 - 2x + 4$
$= -7x + 7$

4. $(4y+5)-(2y-3)=4y+5-2y+3$
$=2y+8$

5. $(-7p-2)-(3p+4)=-7p-2-3p-4$
$=-10p-6$

6. $(4x^2+8x-3)+(9x^2-10x+1)$
$=4x^2+8x-3+9x^2-10x+1$
$=13x^2-2x-2$

7. $(-6m^2-3m-8)-(6m^2+3m-4)$
$=-6m^2-3m-8-6m^2-3m+4$
$=-12m^2-6m-4$

8. $(x^2-x+7)+(-2x^2+4x+6)-(x^2+8)$
$=x^2-x+7-2x^2+4x+6-x^2-8$
$=-2x^2+3x+5$

9. $3x-2(7x^2+2x-1)-(3x^2+8x-2)$
$=3x-14x^2-4x+2-3x^2-8x+2$
$=-17x^2-9x+4$

10. $(-7b)(2b-4)=-14b^2+28b$

11. $3(6x^2-x+1)=18x^2-3x+3$

12. $(-2x^3)(4x^2-3)=-8x^5+6x^3$

13. $(x+5)(x+9)=x^2+9x+5x+45$
$=x^2+14x+45$

14. $(x+3)(x-2)=x^2-2x+3x-6$
$=x^2+x-6$

15. $(3x^3-1)(-4x^4)=-12x^7+4x^4$

16. $(y-3)(4y^2+2y-6)$
$=4y^3+2y^2-6y-12y^2-6y+18$
$=4y^3-10y^2-12y+18$

17. length $=L$, width $=L-3$

18. length of first side $=f$,
length of second side $=f+6$,
length of third side $=2f-2$

19. (a)Number of votes for Lena $=x$,
Number of votes for Jason $=x-3000$
Number of votes for Nhan $=x+5100$
(b) $(x+5100)+x-(x-3000)$
(c) $(x+5100)+x-(x-3000)$
$=x+5100+x-x+3000$
$=x+8100$

20. $3x+12=3(x+4)$

21. $7x^2-14x+21=7(x^2-2x+3)$

22. $2x^2y-6xy^2=2xy(x-3y)$

Cumulative Test for Chapters 1-6

1. 35

2. Saturday

3. $65-20=45$

4. (a) factors $=7,\ x$
product $=-21$
(b) factors $=x,\ y$
product $=z$

5. $5-12(9-10)$
$=5-12(-1)$
$=5-(-12)$
$=17$

6. $7-24\div 6(-2)^2-3$
$=7-24\div 6(4)-3$
$=7-4(4)-3$
$=7-16-3$
$=-9-3=-12$

7. $-6-(-1)=-6+1=-5$

8. $-9-7+(-10)=-16+(-10)=-26$

9. $9(0)(1)(-3)(-2)=0$

10. $0>-72$

11. $-12<-1.2$

12. $|-6|=-(-6)=6,\ |3|=3,\ 6>3\Rightarrow|-6|>|3|$

13. $|8|=8,\ |-8|=-(-8)=8,\ 8=8\Rightarrow|8|=|-8|$

14. $(3x-2)+(-5x+4)=3x-2-5x+4$
$=-2x+2$

15. $(-8y+6)-(6y-7)=-8y+6-6y+7$
$=-14y+13$

16. $(-8x^2+3x-6)-(x^2+5x)$
$=-8x^2+3x-6-x^2-5x$
$=-9x^2-2x-6$

17. $(6x^3+4)+(-2x^3-x)-(5x^2+6x-9)$
$=6x^3+4-2x^3-x-5x^2-6x+9$
$=4x^3-5x^2-7x+13$

18. $4y-3(8y^2+2y-6)-(3y^2+2y-1)$
$=4y-24y^2-6y+18-3y^2-2y+1$
$=-27y^2-4y+19$

19. $(-2a)(3a-4)=-6a^2+8a$

20. $(-6)(4x^2-x+9)=-24x^2+6x-54$

21. $(4y^2+8)(-2y^2)=-8y^4-16y^2$

22. $(x+1)(3x^2-2x+6)$
$=3x^3-2x^2+6x+3x^2-2x+6$
$=3x^3+x^2+4x+6$

23. $(x+2)(x+6)=x^2+6x+2x+12$
$=x^2+8x+12$

24. $(x-5)(x-1)=x^2-x-5x+5$
$=x^2-6x+5$

25. $(2x+7)(x-3)=2x^2-6x+7x-21$
$=2x^2+x-21$

26. $(3x+1)(x+2)=3x^2+6x+x+2$
$=3x^2+7x+2$

27. height of tree $=x$,
height of building $=2x+4$

28. (a) width $=x$, length $=x+6$,
height $=3x$
(b) $3x+x-(x+6)$
(c) $3x+x-(x+6)=4x-x-6=3x-6$

Chapter 7

Pretest Chapter 7

1. $x-9=16 \Rightarrow x=16+9=25$

   check: $25-19\overset{?}{=}16,\ 16=16$

2. $x+8=-3 \Rightarrow x=-3-8=-11$

   check: $-11+8\overset{?}{=}3,\ 3=3$

3. $3x-4+x=-9-3 \Rightarrow 4x=-8 \Rightarrow x=-2$

   check: $3(-2)-4+(-2)\overset{?}{=}-9-3$

   $-12=-12$

4. $\dfrac{y}{4}=3+(-2)^2=3+4=7$

   $y=4(7)=28$

   check: $\dfrac{28}{4}\overset{?}{=}3+(-2)^2,\ 7=7$

5. $\dfrac{-10}{2}=-3(4x)+3x=-12x+3x=-9x$

   $18x=10 \Rightarrow x=\dfrac{10}{18}=\dfrac{5}{9}$

   check: $\dfrac{-10}{2}\overset{?}{=}-3\left(4\cdot\dfrac{5}{9}\right)+3\cdot\dfrac{5}{9},\ -5=-5$

6. $-x=12 \Rightarrow x=-12$

   check: $-(-12)\overset{?}{=}12,\ 12=12$

7. $2z-6=2 \Rightarrow 2z=8 \Rightarrow z=4$

   check: $2(4)-6\overset{?}{=}2,\ 2=2$

8. $-7=-5z-3 \Rightarrow 5z=-3+7=4 \Rightarrow z=\dfrac{4}{5}$

   check: $-7\overset{?}{=}-5\left(\dfrac{4}{5}\right)-3,\ -7=-7$

9. $x+2x=-8-10=-18 \Rightarrow 3x=-18$

   $x=-6$

   check: $-6+2(-6)\overset{?}{=}-8-10,\ -18=-18$

10. $4x-1=8x+6 \Rightarrow 4x=-7 \Rightarrow x=\dfrac{-7}{4}$

    check: $4\left(\dfrac{-7}{4}\right)-1\overset{?}{=}8\left(\dfrac{-7}{4}\right)+6$

    $-8=-8$

11. $2(4x-1)+2x=-42 \Rightarrow 8x-2+2x=-42$

    $10x=-40 \Rightarrow x=-4$

    check: $2(4(-4)-1)+2(-4)\overset{?}{=}-42$

    $-42=-42$

12. $2(x+6)=-3(x-4)+15$

    $2x+12=-3x+12+15 \Rightarrow 5x=15$

    $x=3$

    check: $2(3+6)\overset{?}{=}-3(3-4)+15,\ 18=18$

13. $\left(\dfrac{x}{5}+\dfrac{x}{2}\right)\cdot 10=7\cdot 10 \Rightarrow 2x+5x=70$

    $7x=70 \Rightarrow x=10$

14. $\left(4x+\dfrac{2}{3}\right)\cdot 3=-\dfrac{10}{3}\cdot 3$

    $12x+2=-10 \Rightarrow 12x=-12$

    $x=-1$

15. $\left(\dfrac{x}{3}+x\right)\cdot 3=4(3)=12 \Rightarrow x+3x=12$

    $4x=12$

    $x=3$

16. (a) $P = 40 = 2L + 2W = 2(8 + x) + 2(10)$
$40 = 16 + 2x + 20 \Rightarrow 2x = 4 \Rightarrow x = 2$ ft
(b) $L = 8 + x = 8 + 2 = 10$ ft

17. (a) Henry's tips $= H$,
Amy's tips $= H + 11$
Esther's tips $= H - 6$
(b) $H + (H + 11) + (H - 6) = 95$
(c) $H + (H + 11) + (H - 6) = 95$
$H + H + 11 + H - 6 = 95 \Rightarrow 3H + 5 = 95$
$3H = 90 \Rightarrow H = \$30$, Henry's tips
$H + 11 = \$41$, Amy's tips
$H - 6 = \$24$, Esther's tips
(d) $30 + (30 + 11) + (30 - 6)$
$= 30 + 41 + 24 = 95$

7.1 Exercises

1. To solve the equation $y + 8 = -17$ we add the opposite of $+8$, which is <u>-8</u>, to both sides of the equation.

3. To solve the equation $x - 15 = 82$, we add the number 15 to both sides.

5. Before we solve the equation $7x + 2 - 6x - 9 = 2 - 8$, we must first <u>simplify</u> each side of the equation.

7. To solve the equation $-9y = -36$ we <u>divide by -9</u> on both sides of the equation.

9. To solve the equation $-14 = 3x$, we <u>divide by 3</u> on both sides of the equation.

11. We <u>add -3</u> on both sides of the equation when we solve $14 = x + 3$.

13. $x - 6 = -19 \Rightarrow x = -19 + 6 = -13$
check: $-13 - 6 \stackrel{?}{=} -19$, $-19 = -19$

15. $a + 9 = -1 \Rightarrow a = -1 - 9 = -10$
check: $-10 + 9 \stackrel{?}{=} -1$, $-1 = -1$

17. $-16 = a + 4 \Rightarrow a = -16 - 4 = -20$
check: $-16 \stackrel{?}{=} -20 + 4$, $-16 = -16$

19. $5x + 2 - 4x = 9 \Rightarrow x = 9 - 2 = 7$
check: $5(7) + 2 - 4(7) \stackrel{?}{=} 9$, $9 = 9$

21. $12x - 1 - 11x - 1 = -6 \Rightarrow x - 2 = -6$
$x = -4$
check: $12(-4) - 1 - 11(-4) - 1 \stackrel{?}{=} -6$
$-6 = -6$

23. $12 - 16 = 4y + 6 - 3y \Rightarrow -4 = y + 6$
$y = -4 - 6 \Rightarrow y = -10$
check: $12 - 16 \stackrel{?}{=} 4(-10) + 6 - 3(-10)$
$-4 = -4$

25. $6x - 6 - 5x + 1 = -2 + 4 \Rightarrow x - 5 = 2$
$x = 2 + 5 = 7$
check: $6(7) - 6 - 5(7) + 1 \stackrel{?}{=} -2 + 4$
$2 = 2$

27. $-16 = \frac{x}{3} \Rightarrow x = -16(3) = -48$
check: $-16 \stackrel{?}{=} \frac{-48}{3}$, $-16 = -16$

29. $14 = \frac{y}{-2} \Rightarrow y = 14(-2) = -28$
check: $14 \stackrel{?}{=} \frac{-28}{-2}$, $14 = 14$

31. $\frac{a}{-4} = -6+1 = -5 \Rightarrow a = -5(-4) = 20$

check: $\frac{20}{-4} \stackrel{?}{=} -6+1,\ -5 = -5$

33. $\frac{x}{4} = -8+6 = -2 \Rightarrow x = 4(-2) = -8$

check: $\frac{-8}{4} \stackrel{?}{=} -8+6,\ -2 = -2$

35. $3^2 = \frac{x}{4} \Rightarrow x = 4\left(3^2\right) = 4(9) = 36$

check: $3^2 \stackrel{?}{=} \frac{36}{4},\ 9 = 9$

37. $\frac{y}{-2} = 4+2^2 = 4+4 = 8$

$y = -2(8) = -16$

check: $\frac{-16}{-2} \stackrel{?}{=} 4+2^2,\ 8 = 8$

39. $\frac{x}{7} = -2+3^3 = -2+9 = 7 \Rightarrow x = 7 \cdot 7 = 49$

check: $\frac{49}{7} \stackrel{?}{=} -2+3^2,\ 7 = 7$

41. $3(2x) = 18 \Rightarrow 6x = 18 \Rightarrow x = 3$

check: $3(2(3)) \stackrel{?}{=} 18,\ 18 = 18$

43. $-6 = 2(-5x) \Rightarrow -6 = -10x$

$x = \frac{-6}{-10} = \frac{3}{5}$

check: $-6 \stackrel{?}{=} 2\left(-5\left(\frac{3}{5}\right)\right),\ -6 = -6$

45. $-x = 9 \Rightarrow x = -9$

check: $-(-9) \stackrel{?}{=} 9,\ 9 = 9$

47. $-x = -3 \Rightarrow x = -(-3) = 3$

check: $-(3) \stackrel{?}{=} -3,\ -3 = -3$

49. $8x + 4(4x) = 48 \Rightarrow 8x + 16x = 48$

$24x = 48 \Rightarrow x = 2$

check: $8(2) + 4(4(2)) \stackrel{?}{=} 48,\ 48 = 48$

51. $6(-2x) - 3x = -30 \Rightarrow -12x - 3x = -30$

$-15x = -30 \Rightarrow x = 2$

check: $6(-2(2)) - 3(2) \stackrel{?}{=} -30$

$-30 = -30$

53. $\frac{-10}{2} = 3(4x) + 2x \Rightarrow 12x + 2x = -5$

$14x = -5 \Rightarrow x = \frac{-5}{14}$

check: $\frac{-10}{2} \stackrel{?}{=} 3\left(4 \cdot \frac{-5}{14}\right) + 2 \cdot \frac{-5}{14}$

$-5 = -5$

55. $\frac{21}{3} = 5x + 3(-4x) \Rightarrow 7 = 5x - 12x$

$-7x = 7 \Rightarrow x = -1$

check: $\frac{21}{3} \stackrel{?}{=} 5(-1) + 3(-4(-1)),\ 7 = 7$

57. (a) $2x = -12 \Rightarrow x = \frac{-12}{2} = -6$

(b) $\frac{x}{2} = -12 \Rightarrow x = 2(-12) = -24$

(c) $x - 2 = -12 \Rightarrow x = -12 + 2 = -10$

(d) $x + 2 = -12 \Rightarrow x = -12 - 2 = -14$

59. (a) $y - 10 = 9 \Rightarrow y = 9 + 10 = 19$

(b) $y + 10 = 9 \Rightarrow y = 9 - 10 = -1$

(c) $\frac{y}{-10} = 9 \Rightarrow y = -10(9) = -90$

59. (d) $-10y = 9 \Rightarrow y = \dfrac{9}{-10}$

61. (a) $-2x = 8 \Rightarrow x = \dfrac{8}{-2} = -4$

(b) $-2x \cdot \dfrac{1}{-2} = 8 \cdot \dfrac{1}{-2} \Rightarrow x = \dfrac{8}{-2} = -4$

(c) The answers are the same.

(d) Division is defined in terms of multiplication, therefore dividing by -2 is equivalent to multiplying by the reciprocal of -2, which is $\dfrac{1}{-2}$.

63. $\angle x + 30° = 90° \Rightarrow \angle x = 90° - 30° = 60°$

Cumulative Review Problems

65. $66 = 2 \cdot 3 \cdot 11$

67. $210 = 2 \cdot 3 \cdot 5 \cdot 7$

69. $94{,}500{,}000 - 91{,}400{,}000$
$= 3{,}100{,}000$ mi

7.2 Exercises

1. If 1 is added to double a number, the result is 5. What is the number?

3.

$$\begin{array}{r} 3x+6=5x+9 \\ +\ \boxed{-3x}\quad \boxed{-3x} \\ \hline 0+6=2x+9 \\ +\ \boxed{-9}\quad \boxed{-9} \\ \hline -3=2x+0 \end{array}$$

$$\frac{-3}{\boxed{2}} = \frac{2x}{\boxed{2}}$$

$$\frac{-3}{2} = x$$

5.

$$\begin{array}{r} 6-2x=5-9x \\ +\ \boxed{+9x}\quad \boxed{+9x} \\ \hline 6+7x=5+0 \\ \boxed{-6}\quad \boxed{-6} \\ \hline 0+7x=-1 \end{array}$$

$$\frac{7x}{\boxed{7}} = \frac{-1}{\boxed{7}}$$

$$x = -\frac{1}{7}$$

7. $3x + 9 = 27 \Rightarrow 3x = 18 \Rightarrow x = 6$

check: $3(6) + 9 \stackrel{?}{=} 27,\ 27 = 27$

9. $5x - 10 = 25 \Rightarrow 5x = 35 \Rightarrow x = 7$

check: $5(7) - 10 \stackrel{?}{=} 25,\ 25 = 25$

11. $18 = 4x + 10 \Rightarrow 4x = 8 \Rightarrow x = 2$

check: $18 \stackrel{?}{=} 4(2) + 10,\ 18 = 18$

13. $5x - 1 = 16 \Rightarrow 5x = 17 \Rightarrow x = \dfrac{17}{5} = 3\dfrac{2}{5}$

check: $5\left(\dfrac{17}{5}\right) - 1 \stackrel{?}{=} 16,\ 16 = 16$

15. $-4y + 7 = 63 \Rightarrow -4y = 56 \Rightarrow y = -14$

check: $-4(-14) + 7 \stackrel{?}{=} 63,\ 63 = 63$

17. $-6m - 10 = 88 \Rightarrow -6m = 98 \Rightarrow m = -\dfrac{49}{3}$

check: $-6\left(-\dfrac{49}{3}\right) - 10 \stackrel{?}{=} 88,\ 88 = 88$

19. $40 = -2x - 10 \Rightarrow 50 = -2x \Rightarrow x = -25$

check: $40 \stackrel{?}{=} -2(-25) - 10,\ 40 = 40$

21. $-5y-7=-1 \Rightarrow -5y=6 \Rightarrow y=-\frac{6}{5}$

check: $-5\left(-\frac{6}{5}\right)-7\stackrel{?}{=}-1,\ -1=-1$

23. $6=4-2x \Rightarrow 2=-2x \Rightarrow x=-1$

check: $6\stackrel{?}{=}4-2(-1),\ 6=6$

25. $-3=6-3y \Rightarrow -9=-3y \Rightarrow y=3$

check: $-3\stackrel{?}{=}6-3(3),\ -3=-3$

27. $8y+6-2y=18 \Rightarrow 6y=12 \Rightarrow y=2$

29. $9x-2+2x=6 \Rightarrow 11x=8 \Rightarrow x=\frac{8}{11}$

31. $7x+8-8x=11 \Rightarrow -x=3 \Rightarrow x=-3$

33. $15x=9x+30 \Rightarrow 6x=30 \Rightarrow x=5$

35. $2x=-12x+5 \Rightarrow 14x=5 \Rightarrow x=\frac{5}{14}$

37. $8x+2=5x-4 \Rightarrow 3x=-6 \Rightarrow x=-2$

39. $11x+20=12x+2 \Rightarrow -x=-18 \Rightarrow x=18$

41. $4x-24=6x-8 \Rightarrow -2x=16 \Rightarrow x=-8$

43. $-2+y+8=3y+12 \Rightarrow y+6=3y+12$

$-2y=6 \Rightarrow y=-3$

45. $13y+9-2y=6y-8 \Rightarrow 11y+9=6y-8$

$5y=-17 \Rightarrow y=-\frac{17}{5}$

47. $-9-3x+8=-6x+2-3x$

$-3x-1=-9x+2 \Rightarrow 6x=3 \Rightarrow x=\frac{3}{6}=\frac{1}{2}$

49. $2x-6+8x+13=-2-5x+6$

$10x+7=-5x+4 \Rightarrow 15x=-3 \Rightarrow x=-\frac{1}{5}$

51. $-2y+6=12 \Rightarrow -2y=6 \Rightarrow y=-3$

53. $3x-1=14 \Rightarrow 3x=15 \Rightarrow x=5$

55. $5x+5-2x=15 \Rightarrow 3x=10 \Rightarrow x=\frac{10}{3}$

57. $13x=8x+20 \Rightarrow 5x=20 \Rightarrow x=4$

Cumulative Review Problems

59. $-3(x-4)=-3x+12$

61. $-5(-1+x)=5-5x$

63. $\frac{4}{5}\cdot\frac{3}{3}=\frac{12}{15},\ \frac{2}{3}\cdot\frac{5}{5}=\frac{10}{15} \Rightarrow \frac{4}{5}>\frac{2}{3}$

Israel was overcharged.

7.3 Exercises

1. $-5(2x+1)=25 \Rightarrow -10x-5=25$

$-10x=30 \Rightarrow x=-3$

3. $4(3x-1)=12 \Rightarrow 12x-4=12$

$12x=16 \Rightarrow x=\frac{16}{12}=\frac{4}{3}=1\frac{1}{3}$

5. $36=-2(4y-2) \Rightarrow 36=-8y+4$

$-8y=32 \Rightarrow y=-4$

7. $-5(2x+1)+3x=32$

9. $-4(2x+1) = -5-4x$
$$-8x-4 = -5-4x$$
$$-4x = -1$$
$$x = \frac{1}{4}$$

11. $-2(5y-1)+7y = -1$
$$-10y+2+7y = -1$$
$$-3y = -3$$
$$y = 1$$

13. $4(-6x+2)-6x = 68$
$$-24x+8-6x = 68$$
$$-30x = 60 \Rightarrow x = -2$$

15. $3(y-4)+6(y+1) = 57$
$$3y-12+6y+6 = 57$$
$$9y-6 = 57$$
$$9y = 63$$
$$y = 7$$

17. $2(x-1)+4(x+2) = 18$
$$2x-2+4x+8 = 18 \Rightarrow 6x+6 = 18$$
$$6x = 12 \Rightarrow x = 2$$

19. $10 = -2(x-2)+6(x-1)$
$$10 = -2x+4+6x-6 \Rightarrow 10 = 4x-2$$
$$4x = 12 \Rightarrow x = 3$$

21. $6(x-4) = -9(x+1)+10$
$$6x-24 = -9x-9+10$$
$$15x = 25 \Rightarrow x = \frac{25}{15} = \frac{5}{3}$$

23. $2(y+3) = -6(y-1)-7$
$$2y+6 = -6y+6-7$$
$$8y = -7 \Rightarrow y = -\frac{7}{8}$$

25. $6(x-1)+2(x+1) = 10-3x$
$$6x-6+2x+2 = 10-3x$$
$$8x-4 = 10-3x$$
$$11x = 14 \Rightarrow x = \frac{14}{11}$$

27. $(-6x^2+4x-1)+(6x^2+9) = 12$
$$-6x^2+4x-1+6x^2+9 = 12$$
$$4x+8 = 12$$
$$4x = 4$$
$$x = 1$$

29. $(5x^2+2x+3)-(5x^2+9) = 4x+1$
$$5x^2+2x+3-5x^2-9 = 4x+1$$
$$2x-6 = 4x+1$$
$$2x = -7$$
$$x = -\frac{7}{2}$$

31. $(x^2+3x-1)-(2x+1) = x^2-5$
$$x^2+3x-1-2x-1 = x^2-5$$
$$x-2 = -5 \Rightarrow x = -3$$

33. $(2x^2-6x-3)-(x^2+4) = x^2+6$
$$2x^2-6x-3-x^2-4 = x^2+6$$
$$x^2-6x-7 = x^2+6$$
$$-6x-7 = 6$$
$$-6x = 13 \Rightarrow x = -\frac{13}{6}$$

35. $(2x+9)+(3x-2)-(5x+1)$
$$= 2(x-6)-(x-1)$$
$$2x+9+3x-2-5x-1 = 2x-12-x+1$$
$$6 = x-11 \Rightarrow x = 17 \neq 19$$
19 is not a solution

Cumulative Review Problems

37. $3=3,\ 4=2^2,\ 2=2,\ \text{LCD}=2^2\cdot 3=12$

39. $2x=2x,\ x=x,\ \text{LCD}=2x$

41. $\frac{2}{5}\cdot 925=370$ pages

7.4 Exercises

1. $\frac{x}{2}\cdot 4+\frac{x}{4}\cdot 4=12\cdot 4\Rightarrow 2x+x=48$

$3x=48$

$x=16$

check: $\frac{16}{2}+\frac{16}{4}\stackrel{?}{=}12,\ 12=12$

3. $\frac{x}{6}\cdot 12+\frac{x}{4}\cdot 12=5\cdot 12\Rightarrow 2x+3x=60$

$5x=60$

$x=12$

check: $\frac{12}{6}+\frac{12}{4}\stackrel{?}{=}5,\ 5=5$

5. $\frac{x}{2}\cdot 14-\frac{x}{7}\cdot 14=10\cdot 14\Rightarrow 7x-2x=140$

$5x=140$

$x=28$

check: $\frac{28}{2}-\frac{28}{7}\stackrel{?}{=}10,\ 10=10$

7. $3x\cdot 6+\frac{2}{3}\cdot 6=\frac{9}{2}\cdot 6\Rightarrow 18x+4=27$

$18x=23$

$x=\frac{23}{18}$

check: $3\cdot\frac{23}{18}+\frac{2}{3}\stackrel{?}{=}\frac{9}{2},\ \frac{9}{2}=\frac{9}{2}$

9. $5x\cdot 8+\frac{1}{8}\cdot 8=\frac{3}{4}\cdot 8\Rightarrow 40x+1=6$

$40x=5\Rightarrow x=\frac{5}{40}=\frac{1}{8}$

check: $5\cdot\frac{1}{8}+\frac{1}{8}\stackrel{?}{=}\frac{3}{4},\ \frac{3}{4}=\frac{3}{4}$

11. $5x\cdot 8-\frac{1}{2}\cdot 8=\frac{1}{8}\cdot 8\Rightarrow 40x-4=1$

$40x=5\Rightarrow x=\frac{5}{40}=\frac{1}{8}$

check: $5\cdot\frac{1}{8}-\frac{1}{2}\stackrel{?}{=}\frac{1}{8},\ \frac{1}{8}=\frac{1}{8}$

13. $-2x\cdot 14+\frac{1}{2}\cdot 14=\frac{3}{7}\cdot 14\Rightarrow -28x+7=6$

$-28x=-1\Rightarrow x=\frac{-1}{-28}=\frac{1}{28}$

check: $-2\cdot\frac{1}{28}+\frac{1}{2}\stackrel{?}{=}\frac{3}{7},\ \frac{3}{7}=\frac{3}{7}$

15. $-4x\cdot 6+\frac{2}{3}\cdot 6=\frac{1}{6}\cdot 6\Rightarrow -24x+4=1$

$-24x=-3\Rightarrow x=\frac{-3}{-24}=\frac{1}{8}$

check: $-4\cdot\frac{1}{8}+\frac{2}{3}\stackrel{?}{=}\frac{1}{6},\ \frac{1}{6}=\frac{1}{6}$

17. $\frac{x}{5}+x=6\Rightarrow\frac{6x}{5}=6\Rightarrow 6x=30\Rightarrow x=5$

check: $\frac{5}{5}+5\stackrel{?}{=}6,\ 6=6$

19. $\frac{x}{3}+x=6\Rightarrow\frac{4x}{3}=6\Rightarrow 4x=18\Rightarrow x=\frac{9}{2}$

check: $\frac{\frac{9}{2}}{3}+\frac{9}{2}\stackrel{?}{=}6,\ 6=6$

21. $\frac{x}{2}+x=3 \Rightarrow x+2x=6 \Rightarrow 3x=6$

$x=2$

check: $\frac{2}{2}+2\overset{?}{=}3,\ 3=3$

23. $\frac{x}{4}-3x=3 \Rightarrow x-12x=12$

$-11x=12 \Rightarrow x=-\frac{12}{11}$

check: $\frac{-\frac{12}{11}}{4}-3\left(-\frac{12}{11}\right)\overset{?}{=}3,\ 3=3$

25. $\frac{5}{2}\cdot 6+\frac{x}{3}\cdot 6=\frac{1}{6}\cdot 6 \Rightarrow 15+2x=1$

$2x=-14 \Rightarrow x=-7$

check: $\frac{5}{2}+\frac{-7}{3}\overset{?}{=}\frac{1}{6},\ \frac{1}{6}=\frac{1}{6}$

27. $\frac{3}{2}\cdot 10+\frac{x}{10}\cdot 10=\frac{1}{5}\cdot 10 \Rightarrow 15+x=2$

$x=-13$

check: $\frac{3}{2}+\frac{-13}{10}\overset{?}{=}\frac{1}{5},\ \frac{1}{5}=\frac{1}{5}$

29. $\frac{x}{2}\cdot 6-\frac{2}{6}\cdot 6=-\frac{5}{6}\cdot 6 \Rightarrow 3x-2=-5$

$3x=-3 \Rightarrow x=-1$

check: $\frac{-1}{2}-\frac{2}{6}\overset{?}{=}-\frac{5}{6},\ -\frac{5}{6}=-\frac{5}{6}$

31. $x+\frac{2}{3}+2=\frac{3}{2}+\frac{1}{4}$

$12x+8+24=18+3 \Rightarrow 12x+32=21$

$12x=-11 \Rightarrow x=-\frac{11}{12}$

31. check: $-\frac{11}{12}+\frac{2}{3}+2\overset{?}{=}\frac{3}{2}+\frac{1}{4},\ \frac{7}{4}=\frac{7}{4}$

33. $4+\frac{6}{x}+\frac{2}{5}=\frac{3}{2x} \Rightarrow 40x+60+4x=15$

$44x=-45 \Rightarrow x=-\frac{45}{44}$

check: $4+\frac{6}{-\frac{45}{44}}+\frac{2}{5}\overset{?}{=}\frac{3}{2\left(-\frac{45}{44}\right)},\ \frac{22}{15}=\frac{22}{15}$

Cumulative Review Problems

35. Six more than twice a number: $6+2x$

37. The sum of 4 and *x*: $4+x$

39. $1-\frac{1}{3}-\frac{1}{4}=\frac{12-4-3}{12}=\frac{5}{12}$ of restraurant

7.5 Exercises

1. (a) $2n+5=15$
(b) $2n+5=15 \Rightarrow 2n=10 \Rightarrow n=5$

3. (a) $3n-4=5$
(b) $3n-4=5 \Rightarrow 3n=9 \Rightarrow n=3$

5. (a) $2(5+n)=12$
(b) $2(5+n)=12 \Rightarrow 10+2n=12$
$2n=2 \Rightarrow n=1$

7. $P=60=2L+2W=2(x+15)+2(12)$
$2x+30+24=60 \Rightarrow 2x=6 \Rightarrow x=3$ m

9. $P=x+(x+2)+(x+1)=12$
$3x+3=12 \Rightarrow 3x=9 \Rightarrow x=3$
First side $= x = 3$ cm
second side $= x+1 = 4$ cm
third side $= x+2 = 5$ cm

11. $A = 90 = LW = 2x(15) \Rightarrow 30x = 90$
$x = 3$ in.

13. (a) $P = 70 = 2(15) + 2(x+12)$
$2x + 24 + 30 = 70 \Rightarrow 2x = 16 \Rightarrow x = 8$ ft
(b) width $= x + 12 = 8 + 12 = 20$ ft

15. (a) $A = (9+x)(7) = 105 \Rightarrow 63 + 7x = 105$
$7x = 42 \Rightarrow x = 6$ ft
(b) $L = 9 + x = 9 + 6 = 15$ ft

17. (a) $J =$ Jerome's salary
$J - 7500 =$ cashier's salary
(b) $J + (J - 7500) = 62{,}000$
(c) $2J - 7500 = 62{,}000 \Rightarrow 2J = 69{,}500$
$J = \$34{,}750$, Jerome's salary
$J - 7500 = \$27{,}250$, cashier's salary
(d) $34{,}750 + (34{,}750 - 7500) \stackrel{?}{=} 62{,}000$
$62{,}000 = 62{,}000$

19. (a) $x =$ miles he drove the second day
$x + 85 =$ miles he drove the first day
(b) $x + (x + 85) = 385$
(c) $x + (x + 85) = 385 \Rightarrow 2x = 300$
$x = 150$ mi the second day
$x + 85 = 235$ mi the first day

21. (a) $s =$ flight time of second flight
$\frac{1}{2} \cdot s =$ flight time of first flight
(b) $s + \left(\frac{1}{2} \cdot s\right) = 15$
(c) $s + \left(\frac{1}{2} \cdot s\right) = 15 \Rightarrow \frac{3}{2} \cdot s = 15$
$s = 10$ hr, time of second flight
$\frac{1}{2} \cdot s = 5$ hr, time of first flight

23. (a) $L =$ first side of triangle
$2L =$ second side of triangle
$L + 12 =$ third side of triangle
(b) $L + 2L + (L + 12) = 120$
(c) $4L = 108 \Rightarrow L = 27$ m, first side
$2L = 54$ m, second side
$L + 12 = 39$ m, third side

25. (a) $W =$ width, $3W - 2 =$ length
(b) $2W + 2(3W - 2) = 68$
(c) $2W + 6W - 4 = 68 \Rightarrow 8W = 72$
$W = 9$ m, width
$3W - 2 = 25$ m, length

27. (a) $x =$ students living off campus
$x + 115 =$ students living on campus
$x - 55 =$ students living at home
(b) $x + (x + 115) + (x - 55) = 1704$
(c) $3x + 60 = 1704 \Rightarrow 3x = 1644$
$x = \frac{1644}{3} = 548$ students off campus
$x + 115 = 663$ students on campus
$x - 55 = 493$ students at home

29. $P = x^2 = 3x \Rightarrow x^2 - 3x = 0$
$x(x-3) = 0 \Rightarrow x = 0$ or $x - 3 = 0$
$x - 3 = 0 \Rightarrow x = 3$ ft

Cumulative Review Problems

31. $2459 \times 120 = 295{,}080$

33. $\frac{44}{7} = 6\frac{2}{7}$

35. $L = 4 \cdot 8\frac{1}{4} + 3 \cdot \frac{3}{4} = 35\frac{1}{4}$ in.
$W = 3 \cdot 3\frac{3}{4} + 3 \cdot \frac{3}{4} = 13\frac{1}{2}$ in.

Putting Your Skills to Work, Problems for Individual Investigation

1. $C = 750 = 2x + 50 \Rightarrow 2x = 700$
$x = 350$ hinges

2. $R = 3x - 5 = 3(350) - 5 = \$1045$

3. $P = R - C = 1045 - 750 = \$295$

4. $3x - 5 = 2x + 50 \Rightarrow x = 55$ hinges

5. $P = 500 = (3x - 5) - (2x + 50)$
$x - 55 = 500 \Rightarrow x = 555$ hinges

Putting Your Skills to Work, Problems for Group Investigation and Cooperative Study

1. $C = 2x + 50 + 100 = 2x + 150$

2. $R = 2(3x - 5) = 6x - 10$

3. $P = R - C = (6x - 10) - (2x + 150)$

4. (a) $6x - 10 = 2x + 150 \Rightarrow 4x = 160$
$x = 40$ hinges

(b) $40 < 55$, less

(c) Answers may vary. One explanation: since the break-even point is less for Model B, the company will begin to show a profit sooner.

5.

Model A		Model B	
x hinges	Profit	x hinges	Profit
15	–$40	15	–$100
30	–25	30	–40
40	–15	40	0
45	–10	45	20
55	0	55	60

6. (a) There is a loss.

(b) Model B is more profitable because producing 45 Model B yields a profit of $20 while 45 Model A hinges yields a loss of $10.

7. (a) Model A has 0 profit for 55 hinges. Model B has0 profit for 40 hinges.

(b) Yes. The break-even point occurs when the profit is zero.

Chapter 7 Review Problems

1. $x - 6 = 48 \Rightarrow x = 48 + 6 = 54$

check: $54 - 6 \stackrel{?}{=} 48,\ 48 = 48$

2. $x + 9 = -56 \Rightarrow x = -56 - 9 = -65$

check: $-65 + 9 \stackrel{?}{=} -56,\ -56 = -56$

3. $-13 - 20 = 3x + 15 - 2x$
$x + 15 = -33 \Rightarrow x = -48$

check: $-13 - 20 \stackrel{?}{=} 3(-48) + 15 - 2(-48)$
$-33 = -33$

4. $\frac{y}{-2} = -8 + 3^2 = -8 + 9 = 1 \Rightarrow y = -2(1)$
$y = -2$

check: $\frac{-2}{-2} \stackrel{?}{=} -8 + 3^2,\ 1 = 1$

5. $-6 + 2^2 = \frac{x}{-5} \Rightarrow \frac{x}{-5} = -6 + 4 = -2$
$x = -5(-2) = 10$

check: $-6 + 2^2 \stackrel{?}{=} \frac{10}{-5},\ -2 = -2$

6. $\frac{-15}{3} = -2(5x) - 6x,\ -16x = -5,\ x = \frac{5}{16}$

check: $\frac{-15}{3} \stackrel{?}{=} -2(5\left(\frac{5}{16}\right) - 6\left(\frac{5}{16}\right),\ -5 = -5$

7. $-3x+2(4x)=\frac{18}{-2} \Rightarrow -3x+8x=-9$

$5x=-9 \Rightarrow x=\frac{-9}{5}$

check: $-3\left(\frac{-9}{5}\right)+2\left(4\cdot\frac{-9}{5}\right)\stackrel{?}{=}\frac{18}{-2}$

$-9=-9$

8. $-x=42 \Rightarrow x=-42$

check: $-(-42)\stackrel{?}{=}42,\ 42=42$

9. $-y=\frac{1}{2} \Rightarrow y=-\frac{1}{2}$

check: $-\left(-\frac{1}{2}\right)\stackrel{?}{=}\frac{1}{2},\ \frac{1}{2}=\frac{1}{2}$

10. $6x-8=34 \Rightarrow 6x=42 \Rightarrow x=7$

check: $6(7)-8\stackrel{?}{=}34,\ 34=34$

11. $-2y+6=58 \Rightarrow -2y=52 \Rightarrow y=-26$

check: $-2(-26)+6\stackrel{?}{=}58,\ 58=58$

12. $-20=8-7y \Rightarrow -7y=-28 \Rightarrow y=4$

check: $-20\stackrel{?}{=}8-7(4),\ -20=-20$

13. $46=10-4x \Rightarrow 4x=-36 \Rightarrow x=-9$

check: $46\stackrel{?}{=}10-4(-9),\ 46=46$

14. $1-4y=19 \Rightarrow 4y=-18$

$y=\frac{-18}{4}=\frac{-9}{2}$

check: $1-4\left(\frac{-9}{2}\right)\stackrel{?}{=}19,\ 19=19$

15. $8x-7-5x=15 \Rightarrow 3x=22 \Rightarrow x=\frac{22}{3}$

check: $8\left(\frac{22}{3}\right)-7-5\left(\frac{22}{3}\right)\stackrel{?}{=}15,\ 15=15$

16. $-6x=9x+36 \Rightarrow -15x=36$

$x=\frac{36}{-15}=-\frac{12}{5}$

check: $-6\left(-\frac{12}{5}\right)\stackrel{?}{=}9\left(-\frac{12}{5}\right)+36$

$\frac{72}{5}=\frac{72}{5}$

17. $5x=3x+30 \Rightarrow 2x=30 \Rightarrow x=15$

check: $5(15)\stackrel{?}{=}3(15)+30,\ 75=75$

18. $3x-8=5x+6 \Rightarrow 2x=-14 \Rightarrow x=-7$

check: $3(-7)-8\stackrel{?}{=}5(-7)+6,\ -29=-29$

19. $-3+2y+6=-4y+12$

$2y+3=-4y+12 \Rightarrow 6y=9 \Rightarrow y=\frac{9}{6}=\frac{3}{2}$

check: $-3+2\left(\frac{3}{2}\right)+6\stackrel{?}{=}-4\left(\frac{3}{2}\right)+12$

$6=6$

20. $6y-8+2y=-6+9y$

$8y-8=-6+9y \Rightarrow y=-2$

check: $6(-2)-8+2(-2)\stackrel{?}{=}-6+9(-2)$

$-24=-24$

21. $8x-9-5x+18=-3x-2+2x$

$3x+9=-x-2 \Rightarrow 4x=-11 \Rightarrow x=\frac{-11}{4}$

check : $8\left(\frac{-11}{4}\right)-9-5\left(\frac{-11}{4}\right)+18$

$\stackrel{?}{=}-3\left(\frac{-11}{4}\right)-2+2\left(\frac{-11}{4}\right)$

$\frac{3}{4}=\frac{3}{4}$

22. $-3x-2-9x+5=2x+8-7x$
$-12x+3=-5x+8$
$-7x=5 \Rightarrow x=-\frac{5}{7}$
check: $-3\left(-\frac{5}{7}\right)-2-9\left(-\frac{5}{7}\right)+5 \stackrel{?}{=} 2\left(-\frac{5}{7}\right)+8-7\left(-\frac{5}{7}\right)$
$\frac{81}{7}=\frac{81}{7}$

23. $-3(x+5)=21 \Rightarrow -3x-15=21$
$-3x=36 \Rightarrow x=-12$

24. $4(2x+9)+5x=-3 \Rightarrow 8x+36+5x=-3$
$13x=-39 \Rightarrow x=-3$

25. $7(x+1)+3(x+1)=-10$
$7x+7+3x+3=-10 \Rightarrow 10x+10=-10$
$10x=-20 \Rightarrow x=-2$

26. $-3(x-7)=5(x+6)-10$
$-3x+21=5x+30-10=5x+20$
$-8x=-1 \Rightarrow x=\frac{1}{8}$

27. $(9y^2+8y-2)-(9y^2+5y-1)=14$
$9y^2+8y-2-9y^2-5y+1=14$
$3y-1=14 \Rightarrow 3y=15 \Rightarrow y=5$

28. $(5x^2+x-2)-(5x^2-5)=6x+9$
$5x^2+x-2-5x^2+5=6x+9$
$x+3=6x+9 \Rightarrow -5x=6 \Rightarrow x=-\frac{6}{5}$

29. $\frac{x}{3}\cdot 12+\frac{x}{4}\cdot 12=7\cdot 12 \Rightarrow 4x+3x=84$
$7x=84 \Rightarrow x=12$

30. $2x-\frac{3}{4}=\frac{1}{2} \Rightarrow 8x-3=2 \Rightarrow 8x=5$
$x=\frac{5}{8}$

31. $y+\frac{y}{9}=-10 \Rightarrow 9y+y=-90$
$10y=-90 \Rightarrow y=-9$

32. $2x-\frac{3}{4}=\frac{1}{3} \Rightarrow 24x-9=4 \Rightarrow 24x=13$
$x=\frac{13}{24}$

33. $2y+\frac{y}{-4}=7 \Rightarrow -8y+y=-28$
$-7y=-28 \Rightarrow y=4$

34. $\frac{x}{5}-\frac{x}{2}=6 \Rightarrow 2x-5x=60$
$-3x=60 \Rightarrow x=-20$

35. $-\frac{1}{3}=3y+\frac{1}{2} \Rightarrow -2=18y+3$
$18y=-5 \Rightarrow y=-\frac{5}{18}$

36. (a) $2n+4=16$
(b) $2n=12 \Rightarrow n=6$

37. (a) $2(16)+2W=54$
(b) $32+2W=54 \Rightarrow 2W=22$
$W=11$ ft

38. (a) $P=56=2(15+x)+2(8)$
$30+2x+16=56 \Rightarrow 2x=10 \Rightarrow x=5$ ft
(b) $L=15+x=15+5=20$ ft

39. (a) J = Jamie's salary
$J-8500$ = cashier's salary
(b) $J+(J-8500)=59{,}000$

39. (c) $2J = 67{,}500$

$J = \$33{,}750$, Jamie's salary

$J - 8500 = \$25{,}250$, cashier's salary

(d) $33{,}750 + (33{,}750 - 8500) \stackrel{?}{=} 59{,}000$

$59{,}000 = 59{,}000$

40. (a) $L =$ fall students

$L + 85 =$ spring students

$L - 65 =$ summer students

(b) $L + (L + 85) + (L - 65) = 491$

(c) $3L + 20 = 491 \Rightarrow 3L = 471$

$L = 157$ fall students

$L + 85 = 242$ spring students

$L - 65 = 92$ summer students

(d) $157 - (157 + 85) + (157 - 65) \stackrel{?}{=} 491$

$491 = 491$

Chapter 7 Test

1. $-6x = 72 \Rightarrow x = \dfrac{72}{-6} = -12$

check: $-6(-12) \stackrel{?}{=} 72,\ 72 = 72$

2. $7x + 3 - 6x = -4 \Rightarrow x = -4 - 3 = -7$

check: $7(-7) + 3 - 6(-7) \stackrel{?}{=} -4,\ -4 = -4$

3. $\dfrac{a}{-3} = -4 + 6 = 2 \Rightarrow a = -3(2) = -6$

check: $\dfrac{-6}{-3} \stackrel{?}{=} -4 + 6,\ 2 = 2$

4. $5(2x) = -30 \Rightarrow 10x = -30 \Rightarrow x = -3$

check: $5(2(-3)) \stackrel{?}{=} -30,\ -30 = -30$

5. $5x - 11 = -1 \Rightarrow 5x = 10 \Rightarrow x = 2$

check: $5(2) - 11 \stackrel{?}{=} -1,\ -1 = -1$

6. $\dfrac{x}{5} = -2 + 4^2 = -2 + 16 = 14$

$x = 5(14) = 70$

check: $\dfrac{70}{5} \stackrel{?}{=} -2 + 4^2,\ 14 = 14$

7. $4(-2y) + 5y = 12 \Rightarrow -8y + 5y = 12$

$-3y = 12 \Rightarrow y = -4$

check: $4(-2(-4)) + 5(-4) \stackrel{?}{=} 12,\ 12 = 12$

8. $12y + 6 - 11y - 1 = -8 + 10 = 2$

$y + 5 = 2 \Rightarrow y = -3$

check: $12(-3) + 6 - 11(-3) - 1 \stackrel{?}{=} -8 + 10$

$2 = 2$

9. $-3 = 7 - 4y \Rightarrow 4y = 10 \Rightarrow y = \dfrac{5}{2}$

check: $-3 \stackrel{?}{=} 7 - 4\left(\dfrac{5}{2}\right),\ -3 = -3$

10. $6x + 4 - 9x = 15 \Rightarrow -3x = 11 \Rightarrow x = -\dfrac{11}{3}$

check: $6\left(-\dfrac{11}{3}\right) + 4 - 9\left(-\dfrac{11}{3}\right) \stackrel{?}{=} 15$

$15 = 15$

11. $14x = -2x + 16 \Rightarrow 16x = 16 \Rightarrow x = 1$

check: $14(1) \stackrel{?}{=} -2(1) + 16,\ 14 = 14$

12. $2x + 4(3x - 6) = 4 - (6x + 2)$

$2x + 12x - 24 = 4 - 6x - 2$

$14x - 24 = -6x + 2 \Rightarrow 20x = 26$

$x = \dfrac{26}{20} = \dfrac{13}{10}$

12. check: $2\left(\frac{13}{10}\right)+4\left(3\left(\frac{13}{10}\right)-6\right)$

$\stackrel{?}{=}4-\left(6\left(\frac{13}{10}\right)+2\right)$

$-\frac{29}{5}=-\frac{29}{5}$

13. $3(2x+6)+3x=-27$

$6x+18+3x=-27 \Rightarrow 9x=-45$

$x=-5$

check: $3(2(-5)+6)+3(-5)\stackrel{?}{=}-27$

$-27=-27$

14. $4(x-1)=-6(x+2)+48$

$4x-4=-6x-12+48=-6x+36$

$10x=40 \Rightarrow x=4$

check: $4(4-1)\stackrel{?}{=}-6(4+2)+48,\ 12=12$

15. $(3x^2-2x+1)+(-3x^2-10)=5x+5$

$3x^2-2x+1-3x^2-10=5x+5$

$-2x-9=5x+5 \Rightarrow 7x=-14 \Rightarrow x=-2$

check: $(3(-2)^2-2(-2)+1)$

$+(-3(-2)^2-10)\stackrel{?}{=}5(-2)+5$

$-5=-5$

16. $\frac{x}{5}+\frac{x}{2}=7 \Rightarrow 2x+5x=70$

$7x=70$

$x=10$

17. $2x+\frac{1}{3}=\frac{1}{2} \Rightarrow 12x+4=3$

$12x=-1$

$x=-\frac{1}{12}$

18. $\frac{x}{6}+x=14 \Rightarrow \frac{\cancel{7}}{6}x=\overset{2}{\cancel{14}} \Rightarrow x=6(2)=12$

19. $s=$ length of second side

$s+2=$ length of first side

$3s=$ length of third side

20. $s+(s+2)+3s=42$

21. $5s+2=42 \Rightarrow 5s=40 \Rightarrow s=8$

$s=8$ ft, second side

$s+2=10$ ft, first side

$3s=24$ ft, third side

22. $A=$ Anna's annual salary

$A-4000=$ clerk's annual salary

23. $A+(A-4000)=61{,}200$

24. $2A-4000=61{,}200$

25. $2A=65{,}200$

$A=\$32{,}600$, Anna's annual salary

$A-4000=\$28{,}600$, clerk's salary

Cumulative Test for Chapters 1-7

1. $-3+4+(-1)+8=1+7=8$

2. $2-8-6-9=-6-6-9=-12-9=-21$

3. $(-8)(-2)(2)(-1)=16(-2)=-32$

4. $\frac{16}{2}=8$

5. The opposite of 9 is $\underline{-9}$.

6. The opposite of -12 is $\underline{12}$.

7. The absolute value of 9 is 9.

8. The absolute value of -12 is 12.

9. $\frac{15}{2}=7\frac{1}{2}$

10. $6\frac{3}{4}=\frac{4\cdot 6+3}{4}$
$=\frac{27}{4}$

11. $\frac{25}{40}=\frac{5}{8}$

12. $\frac{3}{4}\cdot\frac{1}{4}=\frac{3}{16}$ lb

13. $2\cdot\frac{3}{4}=\frac{3}{2}=1\frac{1}{2}$ lb

14. $\frac{1}{2}\cdot\frac{3}{4}=\frac{3}{8}$ lb

15. $(2x^2+3xy+4)+(-8x^2+3)$
$=2x^2+3xy+4-8x^2+3$
$=-6x^2+3xy+7$

16. $(8x^2-6x+1)-(6x^2+4x-10)$
$=8x^2-6x+1-6x^2-4x+10$
$=2x^2-10x+11$

17. $12xy+4x^2y-8x=4x(3y+xy-2)$

18. $y+10=-2\Rightarrow y=-2-10=-12$
check: $-12+12\stackrel{?}{=}-2,\ -2=-2$

19. $\frac{x}{-3}=2^2+6=4+6=10$
$x=-3(10)$
$x=-30$
check: $\frac{-30}{-3}\stackrel{?}{=}2^2+6,\ 10=10$

20. $-3x+1=-11\Rightarrow -3x=-12\Rightarrow x=4$
check: $-3(4)+1\stackrel{?}{=}-11,\ -11=-11$

21. $-4(3x)=36\Rightarrow 3x=-9\Rightarrow x=-3$
check: $-4(3(-3))\stackrel{?}{=}36,\ 36=36$

22. $\frac{x}{3}+\frac{x}{2}=5\Rightarrow 2x+3x=30\Rightarrow 5x=30$
$x=6$
check: $\frac{6}{3}+\frac{6}{2}\stackrel{?}{=}5,\ 5=5$

23. $-3x+\frac{x}{2}=4\Rightarrow -6x+x=8$
$-5x=8$
$x=-\frac{8}{5}$
check: $-3\left(-\frac{8}{5}\right)+\frac{-\frac{8}{5}}{2}\stackrel{?}{=}4,\ 4=4$

24. $6x=-8-2x+14\Rightarrow 8x=6$
$x=\frac{6}{8}$
$x=\frac{3}{4}$
check: $6\left(\frac{3}{4}\right)\stackrel{?}{=}-8-2\left(\frac{3}{4}\right)+14,\ \frac{9}{2}=\frac{9}{2}$

25. $4(y-1)=6-(y+8)+12$

$4y-4=6-y-8+12$

$5y=14 \Rightarrow y=\frac{14}{5}$

check: $4\left(\frac{14}{5}-1\right)\stackrel{?}{=}6-\left(\frac{14}{5}+8\right)+12$

$\frac{36}{5}=\frac{36}{5}$

26. $s=$ length of first side

$s-3=$ length of second side

$2s=$ length of third side

27. $s+(s-3)+2s=29$

28. $4s-3=29 \Rightarrow 4s=32 \Rightarrow s=8$

$s=8$ in., first side

$s-3=5$ in, second side

$2s=16$ in., third side

29. $J=$ Juan's annual salary

$J-5500=$ cashier's annual salary

30. $J+(J-5500)=52,000$

31. $2J-5500=52,000 \Rightarrow 2J=57,500$

$J=\$28,750,$ Juan's annual salary

$J-5500=\$23,250,$ cashier's salary

Chapter 8

Pretest Chapter 8

1. 18.071: Eighteen and seventy-one thousandths.

2. $4.542 = 4\frac{542}{1000}$

3. $\frac{7}{100} = 0.07$

4. $0.31 < 0.312$

5. $635.073 = 635.1$ to nearest tenth

6.
$$\begin{array}{r} 3.12 \\ +0.48 \\ \hline 3.60 \end{array}$$

7. $14.1-(-2.5) = 14.1+2.5$
$$\begin{array}{r} 14.1 \\ +\ 2.5 \\ \hline 16.6 \end{array}$$

8. $1.2x + 3x - 5.2y = 4.2x - 5.2y$

9. $15+1+16 = 32$ mi

10.
$$\begin{array}{r} 15.00 \\ -13.57 \\ \hline \$1.43 \end{array}$$

11. $(4.12)(0.3) = 1.236$

12. $(0.536)(10^2) = 53.6$

13. $29.95(2) + 0.30(90) = \$86.90$

14. $8.5 \div 4 = 2.125$

15. $8.2 \div 1.25 = -6.56$

16. (a) $27.5 \div 2.5 = 11$ containers
(b) $(4.26)(2.5) = \$10.65$

17. $0.1x + 0.3x = 2.0 \Rightarrow x + 3x = 20$
$4x = 20 \Rightarrow x = 5$

18. $-2.1x = 14.7 \Rightarrow x = \frac{14.7}{-2.1} = -7$

19. $\frac{86}{100} = 86\%$

20. (a) $\frac{3}{8} = 0.375 = 37.5\%$

(b) $35.8\% = 0.358 = \frac{358}{1000} = \frac{179}{500}$

(c) $4.01\% = 0.0401$

21. (a) $n = 18\% \cdot 53 = 0.18 \cdot 53 = 9.54$
(b) $19 = 72\% \cdot n = 0.72n$
$n = \frac{19}{0.72} = 26.3\overline{8} \approx 26.39$
(c) $35 = n \cdot 312$
$n = \frac{35}{312} = 0.112179487179\cdots$
$n \approx 11.22\%$

22. $\frac{5.25}{30.55} = 0.171849427169\cdots \approx 17.18\%$

23. $p = 88,\ b = 195,\ a$ is unknown

24. p is unknown, $b = 21,\ a = 92$

25. 10% of $11,550 \approx 1155$

26. 1% of 125,282 ≈ 1253

27. 5% of $900 = 45$

28. 15% of \$2560 = 0.15(2560) = \$384

8.1 Exercises

1. When we change 9 to 10, we write 0 and carry 1 to the 2, the next place value to the left. Thus the 2 changes to 3.

3. 5.32: five and thirty-two hundredths

5. 0.428: Four hundred twenty-eight thousandths

7. Three hundred twenty-four thousandths: 0.324

9. Fifteen and three hundred forty-six ten thousandths: 15.0346

11. Twenty-five and 54/100

13. One hundred forty-three and 56/100

15. $0.7 = \frac{7}{10}$

17. $3.64 = 3\frac{64}{100}$

19. $0.1743 = \frac{1743}{10,000}$

21. $100.011 = 100\frac{11}{1000}$

23. $\frac{1}{10} = 0.1$

25. $12\frac{37}{1000} = 12.037$

27. $\frac{2}{100} = 0.02$

29. $\frac{1}{1000} = 0.001$

31. $0.426 < 0.429$

33. $0.63 > 0.62$

35. $0.36 < 0.366$

37. $0.7431 > 0.743$

39. $523.7235 = 523.72$, nearest hundredth

41. $43.995 = 44.00$, nearest hundredth

43. $9.0747 = 9.1$, nearest tenth

45. $462.951 = 463.0$, nearest tenth

47. $312.95166 = 312.952$, nearest thousandth

49. $0.063148 = 0.0631$,
nearest ten-thousandth

51. 42.5 million: 43 million, nearest million
34.6 million: 35 million, nearest million

53. \$15.25: \$15, nearest dollar

55. 0.73, $\frac{7}{10}$, 0.071, 0.007, 0.0069

Cumulative Review Problems

57. $-15-(-6) = -15+6 = -9$

59. $-45 \div 9 = -5$

Putting Your Skills to Work, Problems for Individual Investigation

1.

Check Number	Date	Description of Transaction	Payment/ Debit(−)		Deposit/ Credit(+)		1254	32
243	9/2	Manor Apts.	575					
244	9/2	Elec. Co.	23	41				
245	9/2	Gas Co.	15	67			640	24
246	9/3	Jack's Mkt.	125	57			514	67
247	9/4	Clothing	35	85			478	82
	9/5	Deposit			634	51	1113	33
248	9/7	Bookstore	168	96			944	37
	9/7	ATM	100				844	37
249	9/15	Telephone	43	29				
250	9/15	Sport's Emp.	40					
	9/15	Deposit			634	51	1395	59
251	9/20	Jack's Mkt.	78	94				
252	9/20	Visa Gold	125					
	9/20	ATM	100				1091	65
253	9/25	Drugstore	21	56				
254	9/25	Computer	231	45			838	64
	10/1	Deposit			423	62	1262	26

2. \$1262.26 **3.** Plan 2

Putting Your Skills to Work, Problem for Group Investigation and Cooperative Study

1. $\$1262.26 - \$5.25 = \$1257.01$, check book balance after service charge

Check	Amount	
247	35	85
250	40	
252	125	
254	231	45
total	432	30

$$\begin{array}{r} 1265.69 \\ \underline{+423.62} \\ 1689.31 \\ \underline{-432.30} \\ 1257.01 \end{array}$$

8.2 Exercises

1. To add numbers in decimal notation, we <u>line up</u> the decimal points.

3. When subtracting decimals, we place the decimal point in the answer <u>in line</u> with the decimal point in the problem.

5. $\begin{array}{r} 0.34 \\ +5.23 \\ \hline 5.57 \end{array}$

7. $\begin{array}{r} 35.4 \\ +0.8759 \\ \hline 36.2759 \end{array}$

9. $\begin{array}{r} 0.74 \\ +9.32 \\ \hline 10.06 \end{array}$

11. $\begin{array}{r} 73.0 \\ 7.54 \\ +0.483 \\ \hline 81.023 \end{array}$

13. $\begin{array}{r} 73.1 \\ +0.3169 \\ \hline 73.4169 \end{array}$

15. $\begin{array}{r} 25.0 \\ 2.73 \\ +0.423 \\ \hline 28.153 \end{array}$

17. $\begin{array}{r} 53.783 \\ -2.34 \\ \hline 51.443 \end{array}$

19. $\begin{array}{r} 616.78 \\ -\ 3.9 \\ \hline 612.88 \end{array}$

21. $\begin{array}{r} 20.0 \\ -0.16 \\ \hline 19.84 \end{array}$

23. $\begin{array}{r} 12.1 \\ -0.23 \\ \hline 11.87 \end{array}$

25. $\begin{array}{r} -91.13 \\ -14.213 \\ \hline -105.343 \end{array}$

27. $\begin{array}{r} -8.69 \\ -(-4.12) \\ \hline -4.57 \end{array}$

29. (a) $-3.4+(-2.1)=-5.5$
(b) $9.7-(-5.4)=15.1$
(c) $-9.2-4.1=-13.3$

31. $2.3x+3.9x=6.2x$

33. $24.8y-11.3y=13.5y$

35. $3.5x+9.1x-y=12.6x-y$

37. $1.4x+6.2y+3.5x=4.9x+6.2y$

39. $x+2.3\big|_{x=-6.7}=-6.7+2.3=-4.4$

41. $x-5.1=9.33 \Rightarrow x=9.33+5.1=14.43$

43. $y-0.12\big|_{y=7}=7-0.12=6.88$

45. $16.011-n\big|_{n=9.7}=16.011-9.7=6.311$

47. estimate: $1+2+1=4$ in.

49. estimate: $1763-162-61-48=\$1492$

51. $100-72.31=\$27.69$

53. $10.75-10.54=0.21$ sec faster

55. $11.40-10.54=0.86$ sec

57. $3425.723+181.511+26.809$
$=3634.043$

59. $6007.03-381.95=5625.08$

Cumulative Review Problems

61. $(231)(14)=3234$

63. $(19)(-15)=-285$

65. $\frac{2}{3}\cdot\frac{2}{2}+\frac{1}{2}\cdot\frac{3}{3}=\frac{4+3}{6}=\frac{7}{6}=1\frac{1}{6}$ lb

8.3 Exercises

1. If one factor has 3 decimal places and the second factor has 2 decimal places, the product has 5 decimal places.

3. When we divide $4.62\overline{)12.7}$, we rewrite the equivalent division problem $462\overline{)1270}$ and then divide.

5. $0.03\times 0.07\rightarrow 7\times 3=21$
$0.03\times 0.07=0.0021$

7. $0.04\times 0.08\rightarrow 4\times 8=32$
$0.04\times 0.08=0.0032$

9.
$$\begin{array}{r} 7.43\\ \underline{\times 8.3}\\ 2229\\ \underline{5944}\\ 61.669\end{array}$$

11.
$$\begin{array}{r} 15.2\\ \underline{\times 3.1}\\ 152\\ \underline{456}\\ 47.12\end{array}$$

13.
$$\begin{array}{r} 5.23\\ \underline{\times 1.41}\\ 523\\ 2092\\ \underline{523}\\ 7.3743\end{array}$$

15. $(-4.23)(2.7)=-11.421$

17. $(-25)(-0.613)=15.325$

19. $(12.1)(-2.81)=-34.001$

21. $0.1498\times 100=14.98$

23. $85.54\times 10{,}000=855{,}400$

25. $24\times 10^4=240{,}000$

27. $0.2\times 10^4=2000$

29.
$$\begin{array}{r} 0.62\\ 12\overline{)7.44}\\ \underline{72}\\ 24\\ \underline{24}\end{array}$$

31.
$$\begin{array}{r} 2.16 \\ 8\overline{)17.28} \\ \underline{16} \\ 12 \\ \underline{8} \\ 48 \\ \underline{48} \end{array}$$

33.
$$\begin{array}{r} 0.0565 \\ 64\overline{)3.6160} \\ \underline{320} \\ 416 \\ \underline{384} \\ 320 \\ \underline{320} \end{array}$$

35.
$$\begin{array}{r} 3.451 \\ 24\overline{)84.824} \\ \underline{72} \\ 108 \\ \underline{96} \\ 122 \\ \underline{120} \\ 24 \\ \underline{24} \end{array}$$

37.
$$\begin{array}{r} 0.23 \\ 14\overline{)3.25} \\ \underline{28} \\ 45 \\ \underline{42} \\ 3 \end{array}$$

39. $3.6\overline{)-0.2988} \rightarrow$
$$\begin{array}{r} -0.08 \\ 36\overline{)-2.988} \\ \underline{288} \\ 10 \end{array}$$

41. $-1.7\overline{)-20.8} \rightarrow$
$$\begin{array}{r} 12.23 \\ 17\overline{)208.00} \\ \underline{17} \\ 38 \\ \underline{34} \\ 40 \\ \underline{34} \\ 60 \\ \underline{51} \\ 9 \end{array}$$
$\rightarrow 12.24$

43. $0.27\overline{)6.729} \rightarrow$
$$\begin{array}{r} 24.92 \\ 27\overline{)672.90} \\ \underline{54} \\ 132 \\ \underline{108} \\ 249 \\ \underline{243} \\ 60 \\ \underline{54} \\ 6 \end{array}$$

45. $1.8\overline{)5} \rightarrow$
$$\begin{array}{r} 2.7 \\ 18\overline{)50.0} \\ \underline{36} \\ 140 \\ \underline{126} \\ 14 \end{array}$$
$\rightarrow 2.\overline{7}$

47. $1.1\overline{)0.7} \to 11\overline{)7.0} \to 0.\overline{63}$

$$\begin{array}{r} 0.63 \\ 11\overline{)7.0} \\ \underline{66} \\ 40 \\ \underline{33} \\ 7 \end{array}$$

49. $2.2\overline{)11.3} \to 22\overline{)113.000} \to 5.1\overline{36}$

$$\begin{array}{r} 5.136 \\ 22\overline{)113.000} \\ \underline{110} \\ 30 \\ \underline{22} \\ 80 \\ \underline{66} \\ 140 \\ \underline{132} \\ 8 \end{array}$$

51. $6.6\overline{)200} \to 66\overline{)2000.00} \to 3.0\overline{3}$

$$\begin{array}{r} 30.03 \\ 66\overline{)2000.00} \\ \underline{198} \\ 200 \\ \underline{198} \\ 2 \end{array}$$

53. $\frac{11}{6} = 1.8\overline{3} = 1.83$, nearest hundredth

55. $12\frac{2}{15} = 12.1\overline{3} = 12.13$, nearest hundredth

57. $\frac{1}{3} = 0.\overline{3}$

59. $\frac{7}{9} = 0.\overline{7}$

61. (a) $\frac{1}{300} = 0.00\overline{3}$

(b) $\frac{1}{30} = 0.0\overline{3}$

63. $(-225.7)(128.59) = -29{,}022.763$

65. $562.53 \div 13.123$
$= 42.8659605273\cdots$
$= 42.866$, nearest thousandth

67. $\frac{4}{9} = 0.44\cdots,\ \frac{5}{9} = 0.55\cdots$

Cumulative Review Problems

69. $12x = 96 \Rightarrow x = \frac{96}{12} = 8$

71. $x + 45 = 17 \Rightarrow x = 17 - 45 = -28$

8.4 Exercises

1. $x + 3.7 = 9.8 \Rightarrow x = 9.8 - 3.7 = 6.1$

3. $y - 2.5 = 6.95 \Rightarrow y = 6.95 + 2.5 = 9.45$

5. $2.9 + x = 5 \Rightarrow x = 5 - 2.9 = 2.1$

7. $x + 2.5 = -9.6 \Rightarrow x = -9.6 - 2.5 = -12.1$

9. $5.6 + x = -4.8 \Rightarrow x = -4.8 - 5.6 = -10.4$

11. $3.4x = 10.2 \Rightarrow x = \frac{10.2}{3.4} = 3$

13. $5.1x = 25.5 \Rightarrow x = \frac{25.5}{5.1} = 5$

15. $2x = 11.24 \Rightarrow x = \frac{11.24}{2} = 5.62$

17. $-5.6x = -19.04 \Rightarrow x = \dfrac{-19.04}{-5.6} = 3.4$

19. $-5.2x = 26 \Rightarrow x = \dfrac{26}{-5.2} = -5$

21. $3(x+1.4) = 6.9 \Rightarrow 3x + 4.2 = 6.9$

$3x = 2.7 \Rightarrow x = \dfrac{2.7}{3} = 0.9$

23. $2(x-4) = 26.4 \Rightarrow 2x - 8 = 26.4$

$2x = 34.4 \Rightarrow x = \dfrac{34.4}{2} = 17.2$

25. $6x + 10.5 = x + 21 \Rightarrow 5x = 10.5$

$x = \dfrac{10.5}{5} = 2.1$

27. $2(x+3.2) = x + 9.9 \Rightarrow 2x + 6.4 = x + 9.9$

$x = 9.9 - 6.4 = 3.5$

29. $0.3x + 0.2 = 1.7 \Rightarrow 3x + 2 = 17$

$3x = 15 \Rightarrow x = \dfrac{15}{3} = 5$

31. $0.8x + 0.6 = 5.4 \Rightarrow 8x + 6 = 54$

$8x = 48 \Rightarrow x = \dfrac{48}{8} = 6$

33. $0.12x + 1.1 = 1.22 \Rightarrow 12x + 110 = 122$

$12x = 12 \Rightarrow x = 1$

35. $0.15x + 0.23 = 1.43 \Rightarrow 15x + 23 = 143$

$15x = 120 \Rightarrow x = 8$

37. (a) $\dfrac{901 \text{ mi}}{34 \text{ gal}} = 26.5 \text{ mi/gal}$

(b) $\dfrac{1484 \text{ mi}}{26.5 \text{ mi/gal}} = 56 \text{ gal}$

39. $2(24.95) + 3(12.98) + 54.25 = \$143.09$

41. $3(18.95) + 0.12(423 - 200) = \$83.61$

43. (a) $(2.75 \text{ gal})\left(\dfrac{4 \text{ qt}}{\text{gal}}\right)\left(\dfrac{1 \text{ can}}{\text{qt}}\right) = 11 \text{ cans}$

(b) $(2.75 - 0.75) \text{ gal}\left(\dfrac{1 \text{can}}{1 \text{gal}}\right) = 2 \text{ gal cans}$

$0.75 \text{ gal}\left(\dfrac{4 \text{ qt}}{\text{gal}}\right)\left(\dfrac{1 \text{ can}}{1 \text{ qt}}\right) = 3 \text{ qt cans}$

45. (a) $17(0.24) = \$4.08$

(b) $17(0.14) = \$2.38$

(c) $4.08 - 2.38 = \$1.70$

47. $4.95 + 0.07(75) + 0.10(20) = 12.2$

Single rate plan: \$12.20

$0.10(75) + 0.15(20) = 10.5$

No Fee Plan: \$10.50 is the better deal.

49. $0.25q + 0.10(5q) = 4.5 \Rightarrow 0.75q = 4.5$

$q = 6$ quarters, $5q = 30$ dimes

51. $0.05n + 0.10(5n) = 2.20 \Rightarrow 0.55n = 2.20$

$n = 4$ nickels, $5n = 20$ dimes

53. $23.098x = 103.941 \Rightarrow x = \dfrac{23.098}{103.941} = 4.5$

55. $x + 3.0012 = 21.566$

$x = 21.566 - 3.0012 = 18.5648$

57. $a = \dfrac{v}{t} \Rightarrow v = at = 15(3.5) = 52.5 \text{ ft/sec}$

Cumulative Review Problems

59. $\dfrac{4}{5} = 0.8$

61. $\frac{1}{3} = 0.\overline{3}$

63. $25 - 12\frac{1}{3} = 24\frac{3}{3} - 12\frac{1}{3} = 12\frac{2}{3}$ lb

8.5 Exercises

1. To change a percent to a decimal, move the decimal point 2 places to the <u>left</u> and drop the <u>% sign</u>.

3. $\frac{31}{100} = 31\%$

5. $\frac{7}{100} = 7\%$

7. $\frac{63}{100} = 63\%$

9. $\frac{113}{100} = 113\%$

11. $\frac{16}{100} = 16\%$

13. $\frac{0.9}{100} = 0.9\%$

15. $53.8\% = 0.538$

17. $0.0024 = 0.24\%$

19. $2.33\% = 0.0233$

21. $0.03413 = 3.413\%$

23.

Decimal	Percent
0.576	57.6%
0.249	24.9%
0.003	0.3%
1.546	154.6%

25.

Decimal	Percent
3.7	370.0%
0.238	23.8%
0.006	0.6%
12.882	1288.2%

27. $\frac{35}{40} = 0.875 = 87.5\%$

29. $\frac{3}{4}\% = \frac{3}{4} \cdot \frac{1}{100} = \frac{3}{400}$

31. (a) $\frac{14}{40} = 0.35 = 35\%$

(b) $22.3\% = \frac{22.3}{100} = \frac{223}{1000}$

33. $1\frac{1}{4}\% = \frac{\cancel{5}}{4} \cdot \frac{1}{\cancel{100}_{20}} = \frac{1}{80}$

35. $\frac{1}{40} = 0.025 = 2.5\%$

37. $6.1\% = \frac{6.1}{100} = \frac{61}{1000}$

39. $8.1\% = \frac{8.1}{100} = \frac{81}{1000}$

41. $5.5\% = \frac{5.5}{100} = \frac{11}{200}$

43.

Fraction	Decimal	Percent
$\frac{9}{12}$	0.75	75%
$\frac{61}{100}$	0.61	61%
$\frac{7}{250}$	0.028	2.8%
$9\frac{5}{8}$	9.625	962.5%

45.

Fraction	Decimal	Percent
$\frac{8}{15}$	$0.5\overline{3}$	$53.\overline{3}\%$
$3\frac{1}{2}$	3.5	350%
$\frac{1}{800}$	0.00125	$\frac{1}{8}\%$
$5\frac{1}{4}$	5.25	525%

47. $\frac{16}{35} = 0.457142857143\cdots$

$= 45.7142857143\cdots\%$

$= 45.71\%$, nearest hundredth

49. $\frac{4}{9} = 0.4444444\cdots$

$= 44.44444\cdots\%$

$= 44.44\%$, nearest hundredth

51.

Fraction	$\frac{3}{5}$	$\frac{4}{5}$
Percent	60%	80%

53.

Fraction	$4\frac{2}{8}$	$4\frac{3}{8}$
Percent	425%	437.5%

Cumulative Review Problems

55. $2x = 330 \Rightarrow x = \frac{330}{2} = 165$

57. $x = \frac{1}{3} \cdot 69 = 23$

59. $5.1 - 3.7 = 1.4$ million mi^2

8.6 Exercises

1. 35 is much more than $\frac{1}{2}$ of 40.

3. 9 is very close to 10 so it cannot be $\frac{1}{4}$ of 10.

5. $x = 0.32 \times 84 = 26.88$

7. $n = 0.24 \times 145 = 34.8$

9. $n = 0.46 \times 60 = 27.6$

11. $0.15(12.95) = \$1.94$

13. $0.60(650) = 390$ students

15. $70 = p \times 650 \Rightarrow p = \frac{70}{650} = 0.10769\cdots$

$p = 10.77\%$

17. $60 = n \times 30 \Rightarrow n = \frac{60}{30} = 2 = 200\%$

19. $\frac{15}{80} = 0.1875 = 18.75\%$

21. $\frac{2}{5} = 0.4 = 40\%$

23. $56 = 0.70 \times n \Rightarrow n = 80$

25. $24 = 0.40 \times n \Rightarrow n = 60$

27. $10 = 0.05 \times n \Rightarrow n = 200$

29. $8 = 1.25 \times d \Rightarrow d = 6.4$ mi

31. $x + 0.9x = 95 \Rightarrow 1.9x = 95 \Rightarrow x = \$50$

33. $44 = 0.50 \times n \Rightarrow n = 88$

35. $1.25 \times 85 = n \Rightarrow n = 106.25$

37. $n = 0.27 \times 78 \Rightarrow n = 21.06$

39. $53.94 = n \times 87 \Rightarrow n = 0.62 = 62\%$

41. 0.73×103 million $= 75.19$ million

43. $\frac{86}{92} = 0.93478\cdots = 93.478\cdots\% \to 93\%$

45. $34.5\% - 24.5\% = 10\%$

47. $n = 0.673(348.9) = 234.8097$

49. $368 = 0.20n \Rightarrow n = 1840$

51. $0.346(1,400,000) = n = 484,400$

53. $100\% - (13 + 10 + 27 + 33 + 10)\% = 7\%$

55. $0.13(\$1250) = \$162.5$

Cumulative Review Problems

57. $2x + 3 = 13 \Rightarrow 2x = 10 \Rightarrow x = 5$

59. $5x - 3 = 3x + 9 \Rightarrow 2x = 12 \Rightarrow x = 6$

8.7 Exercises

1. $p = 22,\ b = 250,\ a = 55$

3. $p = 95,\ b = 380,\ a = a$

5. $p = 69,\ b = b,\ a = 8230$

7. $p = p,\ b = 90,\ a = 70$

9. $p = p,\ b = 25,\ a = 10$

11. $p = 160,\ b = b,\ a = 400$

13. $\frac{a}{300} = \frac{24}{100} \Rightarrow a = 72$

15. $\frac{a}{30} = \frac{250}{100} \Rightarrow a = 75$

17. $\frac{a}{4000} = \frac{0.6}{100} \Rightarrow a = 24$

19. $\frac{82}{b} = \frac{50}{100} \Rightarrow b = 164$

21. $\frac{90}{b} = \frac{150}{100} \Rightarrow b = 60$

23. $\frac{4000}{b} = \frac{0.8}{100} \Rightarrow b = 500,000$

25. $\frac{70}{280}=\frac{p}{100}\Rightarrow p=25$

27. $\frac{3.5}{140}=\frac{p}{100}\Rightarrow p=2.5$

29. $\frac{90}{5000}=\frac{p}{100}\Rightarrow p=1.8$

31. $\frac{a}{350}=\frac{26}{100}\Rightarrow a=91$

33. $\frac{540}{b}=\frac{180}{100}\Rightarrow b=300$

35. $\frac{75}{400}=\frac{p}{100}\Rightarrow p=18.75;\ 18.75\%$

37. $\frac{a}{650}=\frac{0.2}{100}\Rightarrow a=1.3$

39. $\frac{15.2}{25}=\frac{p}{100}\Rightarrow p=60.8;\ 60.8\%$

41. $\frac{68}{b}=\frac{40}{100}\Rightarrow b=170$

43. $\frac{94.6}{220}=\frac{p}{100}\Rightarrow p=43;\ 43\%$

45. $\frac{a}{380}=\frac{12.5}{100}\Rightarrow a=47.5$

47. $\frac{a}{5600}=\frac{0.05}{100}\Rightarrow a=2.8$

49. $\frac{a}{250}=\frac{80}{100}\Rightarrow a=200$ apts

51. (a) $100\%-60\%=40\%$

51. (b) $\frac{a}{150}=\frac{40}{100}\Rightarrow a=60$ openings

53. (a) $\frac{42.9}{1950}=\frac{p}{100}\Rightarrow p=2.2;\ 2.2\%$
(b) $100\%-2.2\%=97.8\%$

55. $\frac{90}{b}=\frac{75}{100}\Rightarrow b=120$ units

57. $\frac{6}{9}=\frac{p}{100}\Rightarrow p=66.\bar{6};\ 67\%$

59. $0.1925(798)=153.615\rightarrow 153.62$

61. $0.18(0.20(\$3300))=\$118.80$

Cumulative Review Problems

63. $A=LW=7(4)=28\text{ in}^2$

65. $A=s^2=2^2=4\text{ ft}^2$

8.8 Exercises

1. To estimate 10% of a number, we can delete the last digit.

3. We can find 15% by adding 5% and 10%.

5. We can find 6% by adding 5% and 1%.

7. 10% of 801: 80

9. $5\%(801)\approx\frac{1}{2}(80)=40$

11. $20\%(801)\approx 2(80)=160$

13. $1\%(205)\approx 2$

15. $7\%(205) = 5\%(205) + 2\%(205)$
$\approx \frac{1}{2}(20) + 2(2) = 14$

17. $30\%(205) = 3(10\%(205)) \approx 3(20) = 60$

19. $10\%(1007) \approx 100$

21. $15\%(1007) = 10\%(1007) + 5\%(1007)$
$= 10\%(1007) + \frac{1}{2} \cdot 10\%(1007)$
$\approx 100 + \frac{1}{2}(100) = 150$

23. $20\%(1007) = 2(10\%(1007))$
$\approx 2(100) = 200$

25. $1\%(3015) \approx 30$

27. $2\%(3015) = 2(1\%(3015)) \approx 2(30) = 60$

29. $3\%(3015) = 3(1\%(3015) \approx 3(30) = 90$

31. $10\%(320,050) \approx 10\%(320,000)$
$= 32,000$

33. $5\%(320,050) \approx 5\%(320,000)$
$= \frac{1}{2} \cdot 10\%(320,000) = \frac{1}{2} \cdot 32,000$
$= 16,000$

35. $15\%(320,050) \approx 15\%(320,000)$
$= 10\%(320,000) + 5\%(320,000)$
$= 32,000 + \frac{1}{2}(10\%(320,000))$
$= 32,000 + \frac{1}{2}(32,000) = 48,000$

37. $1\%(250,030) \approx 1\%(250,000) = 2500$

39. $4\%(250,030) \approx 4(1\%(250,000))$
$= 4(2500) = 10,000$

41. $8\%(250,030) \approx 8(1\%(250,000))$
$= 8(2500) = 20,000$

43. $5\%(\$19,999) \approx 5(1\%(\$20,000))$
$= 5(\$200) = \$1000$

45. $6\%(\$139,999) \approx 6(1\%(\$140,000))$
$= 6(\$1400) = \$8400$

47. (a) $20 + 200 + 2(50) = \$320$
(b) $40\%(320) \approx 4(10\%(300))$
$= 4(30) = \$120$
(c) $320 - 120 = \$200$

49. $0.60(19.99 + 199 + 2(49.99)) = \$191.38$
$200 - 191.38 = \$8.62$

Cumulative Review Problems

51. $\frac{5}{n} = \frac{7.5}{18} \Rightarrow 7.5n = 90 \Rightarrow n = 12$

53. $1 - \frac{1}{4} - \frac{3}{8} = \frac{8-2-3}{8} = \frac{3}{8}$ of the tank

Putting Your Skills to Work, Problems for Individual Investigation

1. $10\%(\$240,000) = \$24,000$
$5\%(\$240,000) = \frac{1}{2} \cdot 10\%(\$240,000)$
$= \frac{1}{2} \cdot \$24,000 = \$12,000$
$1\%(\$240,000) = \$2400$

2. (a) $5\% + 1\% \rightarrow 12,000 + 2400 = \$14,400$

2. (b) $2\% = 2(1\%) \to 2(2400) = \$4800$

(c) $\$4800 + \$14,400 = \$19,200$

3. $\$240,000 - \$19,200 = \$220,800$

Putting Your Skills to Work, Problems for Group Investigation and Cooperative Study

1. (a) $\$1800 = 1\% \cdot x \Rightarrow x = \$180,000$

(b) $\$240,000 - \$180,000 = \$60,000$

(c) $2\%(\$240,000) = 2(1\%(\$240,000))$
$= 2(\$2400) = \$4800$

(d) $1\frac{1}{2}\%(\$180,000)$

$= 1\%(\$180,000) + \frac{1}{2} \cdot 1\%(\$180,000)$

$= \$1800 + \frac{1}{2} \cdot \$1800 = \$2700$

(e) $\$60,000 + \$4800 + \$2700 = \$67,500$

Sale Price	\$240,000
(a) Loan amount	\$180,000
(b) Down payment	\$60,000
(c) Escrow fee	\$4800
(d) Loan fees	\$2700
(e) Total cash needed	\$67,500

2. (a) $\$160,000 - \$90,000 = \$70,000$

(b) $6\%(\$160,000) = \$9600$

(c) $2\%(\$160,000) = \$3200$

(d) $\$70,000 - \$9600 - \$3200 = \$57,200$

(e) Toward the purchase they have $\$57,200 + \$10,000 = \$67,200$. They need \$67,500 so they are short \$300.

2.

Sale Price of Townhouse	\$160,000
Payoff of existing loan	\$90,000
(a) Funds left after payoff	\$70,000
(b) Real estate fee	\$9600
(c) Escrow fees	\$3200
(d) Funds left for purchase	\$57,200

8.9 Exercises

1. $0.15(2350) = \$352.50$

3. $0.25(13,500) = \$3375$

5. $\frac{379.5}{1725} = 0.22 = 22\%$

7. $1.08(35,500) = \$38,340$

9. $1.15(760) = 874$ people

11. (a) $0.07(3000) = \$210$

(b) $3000 + 210 = \$3210$

13. $I = PRT = 500(0.08)(0.5) = \$20$

15. $1.06x = 900 \Rightarrow x = \$849.06$

17. $\frac{3500 + 3500(0.08)(1)}{12} = \$315$

19. $265 - 0.15(265) = 225.25$ lb

21. $0.05x = 35 \Rightarrow x = \$700$

23. (a) $2(289) + 421 = \$999$

(b) $0.7(999) = \$699.30$

25. (a) $1.097x = 1086 \Rightarrow x = \$989.97$

25. (b) $1.097(1086) = \$1191.34$

27. (a) $0.68(22,437) = 15,257.16$ mi
(b) $15,257.16(0.31) = \$4729.72$

29. (a) $0.03(35)(0.11)(320) = \$36.96$
(b) $0.07(35)(0.11)(320) - 50 = \$36.24$

Cumulative Review Problems

31. $A = bh = 6(4) = 24\ \text{cm}^2$

Chapter 8 Review

1. 6.23: Six and twenty-three hundredths

2. 0.679: Six hundred seventy-nine thousandths

3. 7.0083: Seven and eighty-three ten thousandths

4. Forty-six and 85/100

5. $4.267 = 4\frac{267}{1000}$

6. $43.91 = 43\frac{91}{100}$

7. $32\frac{761}{1000} = 32.761$

8. $54\frac{26}{1000} = 54.026$

9. $0.523 < 0.524$

10. $0.16 < 0.168$

11. $842.8569 = 842.86$, nearest hundredth

12. $406.7809 = 406.781$, nearest thousandth

13. $0.52 + 8.11 = 8.63$

14. $-5.2 + 0.236 = -4.964$

15. $0.588 + 36 + 8.43 = 45.018$

16. $25.98 - 2.33 = 23.65$

17. $-2.12 - 9.67 = -11.79$

18. $-9.355 - 2.48 = -11.835$

19. $(-9.2) + (-5.4) = -14.6$

20. $4.32 - (-6.43) = 10.75$

21. $-7 - (-6.67) = -0.33$

22. To subtract numbers in decimal notation, we line up the decimal points.

23. When adding $85 + 36.5$, we rewrite 85 as 85.0 so that we can line up the decimal points.

24. $x - 9.3\big|_{x=0.6} = 0.6 - 9.3 = -8.7$

25. $y + 17.2\big|_{y=-2.3} = -2.3 + 17.2 = 14.9$

26. $4.6x + 7.2x = 11.8x$

27. $8.6x + 3.9x = 12.5x$

28. $23 + 25 + 21 + 24 = \$93$

29. $100 - 54.56 - 21.06 = \$24.38$

30. $0.091 \times 0.06 = 0.00546$

31. $0.082 \times 0.02 = 0.00164$

32. $5.68 \times 7.21 = 40.9582$

33. $2.62 \times 7.33 = 19.2046$

34. $(3.01)(-41.25) - -124.1625$

35. $(-5.6)(-9.01) = 50.456$

36. If one factor has 2 decimal places and the second factor has 3 decimal places, the product has <u>five</u> decimal places.

37. If one factor has 3 decimal places and the second factor has 4 decimal places, the product has <u>seven</u> decimal places.

38. $0.1249 \times 100 = 12.49$

39. $3.24 \times 1000 = 3240$

40. $41 \times 10^5 = 4{,}100{,}000$

41. $40(8.30) + 9(12.45) = \$444.05$

42. $4500 + 0.12(44{,}322.6 - 36{,}000)$
$= \$5498.71$

43. $8.66 \div 12 = 0.72$

44. $-3.25 \div 5.1 = -0.64$

45. $-8.52 \div -7.2 = 1.18$

46. $16.221 \div 0.33 = 49.1\overline{54}$

47. $13.01 \div 0.33 = 39.\overline{42}$

48. When we divide $7.21\overline{)25.9}$, we rewrite the equivalent division problem

48. $\underline{721\overline{)2590}}$ and then divide.

49. $12\overline{)4.\dot{3}49}$

50. $\frac{86}{8} = 10.75$

51. $9\frac{1}{5} = 9.2$

52. $\frac{13}{9} = 1.\overline{4} = 1.44$, nearest hundredth

53. (a) $38.5 \div 3.5 = 11$ containers
(b) $3.5(6.20) = \$21.70$

54. $x - 2.68 = 8.23 \Rightarrow x = 8.23 + 2.68 = 10.91$

55. $-1.6x = 3.68 \Rightarrow x = \frac{3.68}{-1.6} = -2.3$

56. $2x + 2.4 = 8.7 \Rightarrow x = \frac{8.7 - 2.4}{2} = 3.15$

57. $3x - 2.8 = 2x + 4.2 \Rightarrow x = 4.2 + 2.8 = 7$

58. $0.05(5d) + 0.10d = 14 \Rightarrow 0.35d = 14$
$d = \frac{14}{0.35} = 40$ dimes
$5d = 200$ nickels

59. $\frac{85}{100} = 85\%$

60. $\frac{132}{110} = 1.2 = 120\%$

61. $5.7\% = 0.057$

62. $0.016 = 1.6\%$

63. $124\% = 1.24$

64.

Decimal Form	Percent Form
0.379	37.9%
0.428	42.8%
0.005	0.5%
3.47	347%

65.

Decimal Form	Percent Form
1.2	120%
0.035	3.5%
0.0025	0.25%
0.567	56.7%

66. $\frac{56}{168} = 0.\bar{3} = 33.\bar{3}\% = 33\frac{1}{3}\%$

67. $81.3 = 81\frac{3}{10}$

68.

Fraction Form	Decimal Form	Percent Form
$\frac{3}{8}$	0.375	37.5%
$\frac{18}{25}$	0.72	72%
$\frac{1}{200}$	0.005	$0.5\% = \frac{1}{2}\%$
$7\frac{2}{5}$	7.4	740%

69.

Fraction Form	Decimal Form	Percent Form
$\frac{3}{4}$	0.75	75%
$\frac{14}{25}$	0.56	56%
$\frac{1}{500}$	0.002	$0.2\% = \frac{1}{5}\%$
$3\frac{6}{15}$	3.4	340%

70. $82 = 0.25n \Rightarrow n = 328$

71. $0.30(90) = 27$

72. $15 = p \cdot 300 \Rightarrow p = 0.05 = 5\%$

73. $0.45(120) = 54$

74. $\frac{18}{1.15} \approx \$15.65$

75. $26 = 0.10n \Rightarrow n = 260$

76. $0.07(17{,}000) = \$1190$

77. $\frac{4.25}{32.40} = 0.1311772839506\cdots \approx 13.12\%$

78. $p = 85,\ b = 400,\ a = a$

79. $p = p,\ b = 100,\ a = 20$

80. $9 = 0.45b \Rightarrow b = 20$

81. $22 = p(88) \Rightarrow p = 0.25 = 25\%$

82. (a) The original price of the dishwasher in the advertisement is $\underline{225+75=\$300}$.

(b) $75 = p(300) \Rightarrow p = 0.25 = 25\%$
The dishwasher is marked down $\underline{25\%}$.

83. (a) $100\% - 70\% = 30\%$
(b) $0.30(250) = 75$

84. $10\%(176,001) \approx 10\%(176,000) = 17,600$

85. $1\%(176,001) \approx 1\%(176,000) = 1760$

86. $7\%(176,001) \approx 7\%(176,000)$
$= \frac{1}{2} \cdot 10\%(176,000) + 2 \cdot 1\%(176,000)$
$= 12,320$

87. $20\%(176,001) \approx 2 \cdot 10\%(176,000)$
$= 35,200$

88. $0.18(13,250) = \$2385$

89. $\frac{59,500 - 50,0000}{50,0000} = 0.19 = 19\%$

90. $I = PRT = 7500(0.13)(1) = \$975$

Chapter 8 Test

1. 207.402: Two hundred seven and four hundred two thousandths

2. $0.013 = \frac{13}{1000}$

3. $\frac{51}{100} = 0.51$

4. $0.45 > 0.412$

5. $746.136 = 746.14$, nearest hundredth

6. $12.93 + 0.21 = 13.14$

7. $x + 0.12\big|_{x=-2.07} = -2.07 + 0.12 = -1.95$

8. $3.1x + 2.01y + 1.06x = 4.16x + 2.01y$

9. $18.81 - 6.17 = 12.64$

10. $(-13.2) - (-7.1) = -6.1$

11. $(8.24)(1.2) = 9.888$

12. $(4.72)(10^3) = 4720$

13. $15.75 \div 3.5 = 4.5$

14. $\frac{19}{3} = 6.\overline{3}$

15. $\frac{7}{100} = 7\%$

16. (a) $\frac{76}{95} = 0.8$
(b) $\frac{76}{95} = 0.8 = 80\%$

17. (a) $0.10 = \frac{10}{100} = \frac{1}{10}$
(b) $0.10 = \frac{10}{100} = 10\%$

18. (a) $5\% = \frac{5}{100} = \frac{1}{20}$
(b) $5\% = \frac{5}{100} = 0.05$

19. $0.24(3300) = \$792$

20. $\dfrac{249}{1.2} = \$207.50$

21. $5500 + 0.12(34{,}100.5 - 30{,}000)$
$= \$5992.06$

22. (a) $1.74 - 1.38 = 0.36 \rightarrow 36$ cents more
(b) $3.55 - 3.68 = -0.13 \rightarrow 13$ cents less

23. $\dfrac{4.29 - 3.27}{3.27} \approx 0.3119 = 31.2\%$

24. $0.5x + 0.2x = 2.8 \Rightarrow x = \dfrac{2.8}{0.7} = 4$

25. $2(y + 1.3) = 7.8 \Rightarrow y = \dfrac{7.8}{2} - 1.3 = 2.6$

26. (a) $\dfrac{577.2}{33.1} \approx 17.4$ mi/gal
(b) $33.1(1.27) = \$42.04$

27. $12 = 0.30n \Rightarrow n = \dfrac{12}{0.30} = 40$

28. $5 = p(250) \Rightarrow p = \dfrac{5}{250} = 0.02 = 2\%$

29. $a = 0.20(48) = 9.6$

30. $0.15(10{,}500) = \$1575$

31. $I = PRT = 8200(0.11)(2) = \$1804$

32. One Step: $0.80(850) - 50 = \$630$
One Touch:
$0.85(850) - 100 = \$622.5$, better buy

Cumulative Test for Chapters 1-8

1. $x + 2x + 3xy = 3x + 3xy$

2. $5x - x + 2x = 4x + 2x = 6x$

3. $3 - 6x + 2xy - 6 - 4xy = -3 - 6x - 2xy$

4. $P = 4x = 4(8) = 32$ ft

5. $A = LW = 11(9) = 99$ ft^2

6. $-10{,}000 + 25{,}000 = \$15{,}000$ profit

7. $4x - 2\big|_{x=-1} = 4(-1) - 2 = -6$

8. $4x - 2\big|_{x=10} = 4(10) - 2 = 38$

9. $4x - 2\big|_{x=\frac{1}{2}} = 4\left(\dfrac{1}{2}\right) - 2 = 0$

10. $\dfrac{250}{30} = 8\dfrac{1}{3}$ cal/min

11. $\dfrac{18y}{27y^2} = \dfrac{2}{3y}$

12. $(x^3)^2 = x^{3 \cdot 2} = x^6$

13. $(4xy)^3 = 4^3x^3y^3 = 64x^3y^3$

14. $\dfrac{2}{3} = \dfrac{26}{x} \Rightarrow 2x = 78 \Rightarrow x = 39$

15. $(-3x)(2x - 5) = -6x^2 + 15x$

16. $(-2x + 1) + (5x - 6) = -2x + 1 + 5x - 6$
$= 3x - 5$

17. $3.2x = 19.2 \Rightarrow x = \dfrac{19.2}{3.2} = 6$

18. $0.12x + 0.5x = 1.86 \Rightarrow x = \dfrac{1.86}{0.62} = 3$

19. $0.75 + 0.15(7) = \$1.80$

20. $180.273 = 180.27,$ nearest hundredth

21. $20.17 + 13.59 = 33.76$

22. $27.01 - 5.3 = 21.71$

23. $(9.3)(10^2) = 930$

24. $14.7 \div 4.2 = 3.5$

25. $1.15(15) = \$17.25$

26. (a) $\dfrac{15}{25} = 0.6$

(b) $\dfrac{15}{25} = 0.6 = 60\%$

27. (a) $0.30 = \dfrac{30}{100} = \dfrac{3}{10}$

(b) $0.30 = \dfrac{30}{100} = 30\%$

28. (a) $1\% = \dfrac{1}{100}$

(b) $1\% = \dfrac{1}{100} = 0.01$

Chapter 9

Chapter 9 Pretest

1. \$18,000

2. \$40,000

3. $40,000 - 25,000 = \$15,000$

4. $32,000 - 19,000 = \$13,000$ less

5. $\frac{3+5+2+0+2+4+3}{7} = \frac{19}{7} \approx 2.7$ mi

6. $0, 2, 2, 3, 3, 4, 5 \to$ median = 3 mi

7.

8.

9.

10.

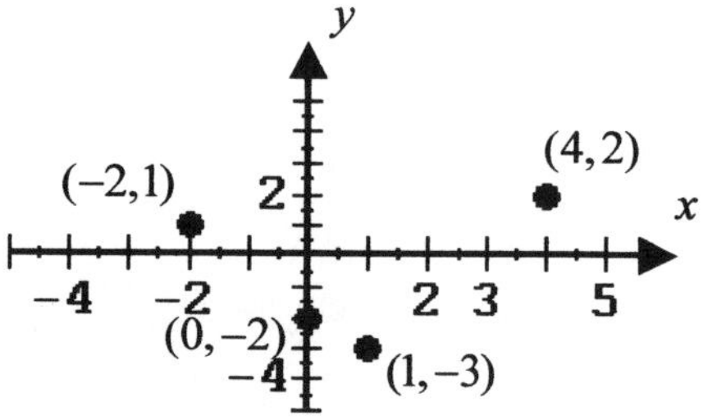

11. (a) (1,3), (0,1), and (−1,−1) are solutions to $y = 2x + 1$.

(b)

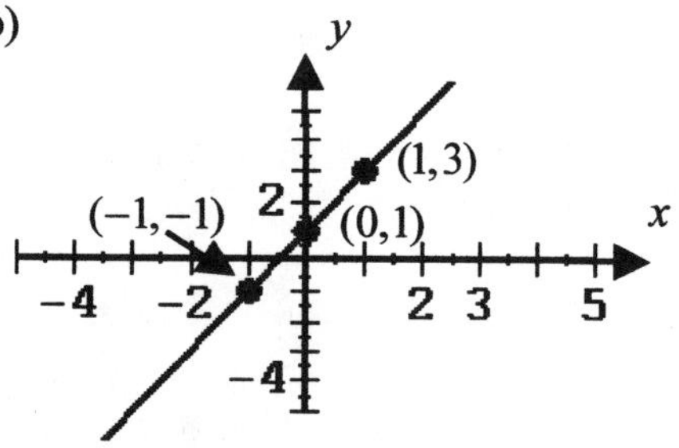

Putting Your Skills to Work, Problems for Individual Investigation

1. $13,440 + 2160 + 3360 + 1920 + 2640 + 480 = \$24,000$

2.

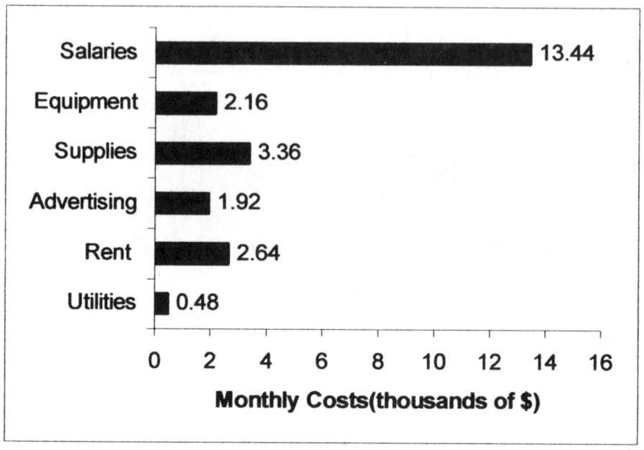

3. Salaries: $\frac{13,440}{24,000} = 0.56 = 56\%$

Equipment: $\frac{2160}{24,000} = 0.09 = 9\%$

Supplies: $\frac{3360}{24,000} = 0.14 = 14\%$

Advertising: $\frac{1920}{24,000} = 0.08 = 8\%$

Rent: $\frac{2640}{24,000} = 0.11 = 11\%$

Utilities: $\frac{480}{24,000} = 0.02 = 2\%$

4.

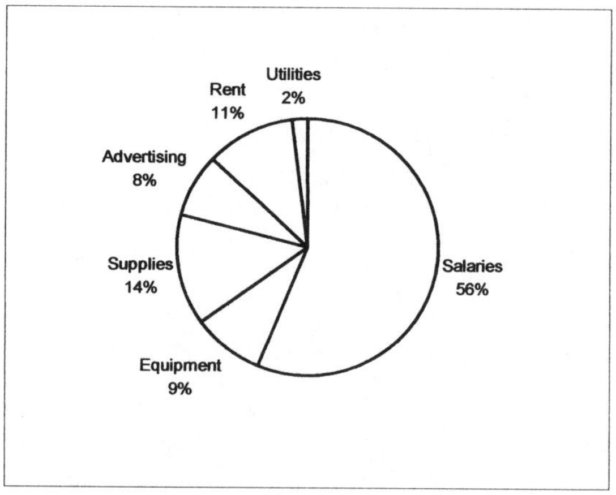

5. $1.04(24{,}000) = \$24{,}960$

Putting Your Skills to Work, Problems for Group Investigation and Cooperative Study

1. 2001 Monthly Production Costs

Salaries	$1.05(13{,}440) = \$14{,}112$
Equipment	$0.40(2160) = \$864$
Supplies	$1.10(3360) = \$3696$
Advertising	$0.79(1920) = \$1516.80$
Rent	$1.05(2640) = \$2772$
Utilities	$480

2. (a) $23,440.80

(b) $24{,}000 - 23{,}440.80 = \$559.20$ less

(c) $\frac{559.20}{24{,}000} = 0.0233 = 2.33\%$ less

3.

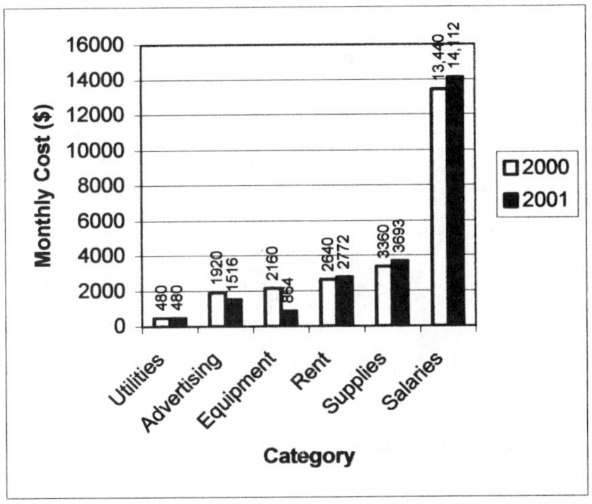

4. (a)

Salaries	$14{,}112 \div 23{,}440 \approx 0.60 = 60\%$
Equipment	$864 \div 23{,}440 \approx 0.04 = 4\%$
Supplies	$3696 \div 23{,}440 \approx 0.16 = 16\%$
Advertising	$1516.8 \div 23{,}440 \approx 0.06 = 6\%$
Rent	$2772 \div 23{,}440 \approx 0.12 = 12\%$
Utilities	$480 \div 23{,}440 \approx 0.02 = 2\%$

4. (b)

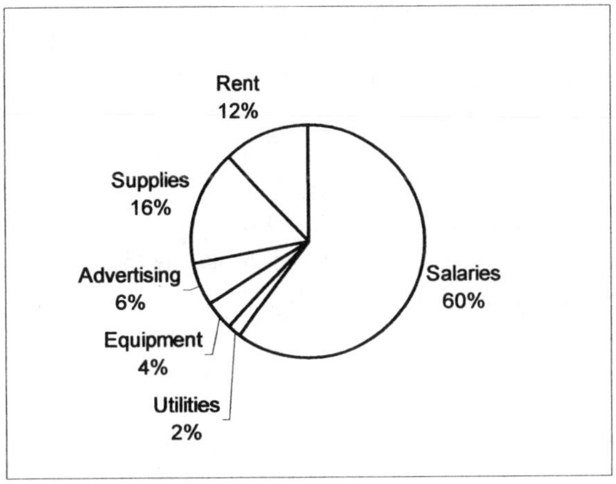

9.1 Exercises

1. $3(50{,}000) = 150{,}000$

3. Systems analysts

5. $9(50{,}000) - 3(50{,}000) = 300{,}000$

7. $13(17) = 221\ \text{people/mi}^2$

9. $10(17) - 4(17) = 102$ more people

11. 31.76%

13. (a) Housing and transportation
(b) $31.76\% + 20.24\% = 52\%$

15. $0.3176(18{,}450) = \$5{,}859.72$

17. 17.1% **19.** $10.4\% + 16.1\% = 26.5\%$

21. $100\% - 17.1\% = 82.9\%$ **23.** 42.7 in.

25. Anchorage, AK and Buffalo, NY

27. $63.3 - 31.9 = 31.4$ in.

29. Boston: $42.7-24.90=17.8$
Denver: $60.7-36.9=23.8$
Flagstaff: $109.1-66.5=42.6$
Great Falls: $63.3-31.9=31.4$
Omaha: $28.7-13.0=15.7$
Anchorage: $68.5-77.4=-8.9$
Buffalo: $93.0-105.1=-12.1$
Boise: $21.6-14.2=7.4$
Flagstaff, AZ with 42.6 in. has the greatest difference.

31. June

33. (a) 4.5 thousand or 4500
(b) increased from 3500 to 4500

35. Between June and July, an increase of 2500.

Month	Inc(+)/Dec(−)
Mar-Apr	$2000-3000=-1000$
Apr-May	$3500-2000=1500$
May-Jun	$4500-3500=1000$
Jun-Jul	$7000-4500=2500$
Jul-Aug	$4000-7000=-3000$

37. 4 in. **39.** April, May, and June

41. $4-1.5=2.5$ in.

43.

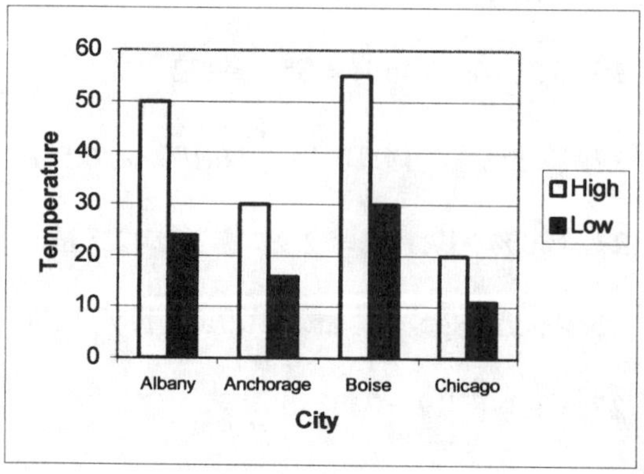

45.

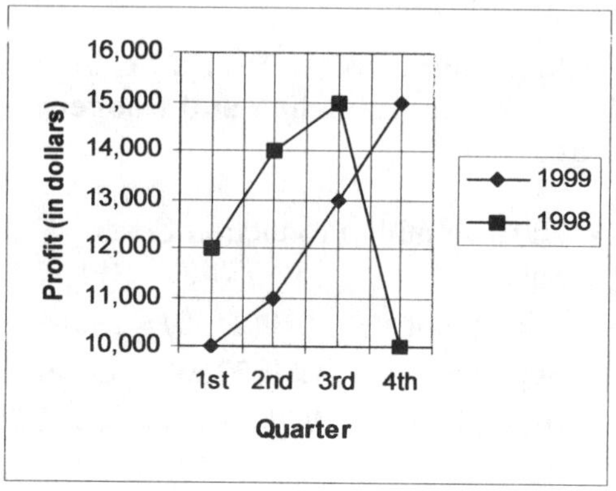

47.

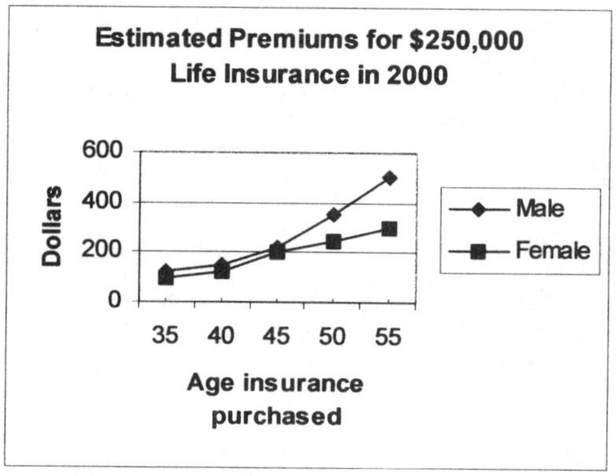

49. $100\%-25.3\%=74.7\%$

51. $0.175(100,000)(\$102)=\$1,785,000$

Cumulative Review Problems

53. $A=LW=14(6)=84\text{ in.}^2$

55. $(2(7)(12)+2(7)(20))\text{yd}^2\cdot\dfrac{1\text{ gal}}{28\text{ yd}^2}$
$=16$ gal

9.2 Exercises

1. $\dfrac{89+92+83+96+99}{5}=91.8$

3. $\dfrac{23+45+63+34+21+42}{6}=38$

5. $\dfrac{189+193+162+102+189}{5}=167$

$167,000

7. $\dfrac{2+3+3+4+2.5+2.5+4}{7}=3\text{ hr}$

9. $\dfrac{6+5+4++5+8}{5}=5.6\text{ sales}$

11. $\dfrac{0+2+3+2+2}{5+4+6+5+4}=\dfrac{9}{24}=0.375$

13. $\dfrac{276+350+391+336}{12+14+17+14}$

$=\dfrac{1353}{57}\approx 23.7\text{ mi/gal}$

15. 22, 36, 45, 47, 48, 50, 58

↑

median = 47

17. 865, 968, 999, 1023, 1052, 1152

↑

median = $(999+1023)\div 2=1011$

19. 0.34, 0.52, 0.58, 0.69, 0.71

↑

median = 0.58

21. $11,600
$15,700
$17,000 ←
$23,500
$26,700
$31,500
median = $(17,000+23,500)\div 2=\$20,500$

23. 12, 24, 31, 40, 62, 108

↑

median = $(31+40)\div 2=35.5$ min

25. $5.99
$7.99
$9.99 ←
$11.99
$13.99
$17.99
median = $(9.99+11.99)\div 2=\$10.99$

27. 22, 36, 36, 37, 44, 48, 53, 60, 64, 71

↑

median = $(44+48)\div 2=46$ actors

29. 59, $\underbrace{60, 60,}_{\text{mode}}$ 65, 68, 72, 80

31. 116, 117, $\underbrace{121, 121,}_{\text{mode}}$ 123, $\underbrace{150, 150}_{\text{mode}}$

33. $\underbrace{\$249, \$249,}_{\text{mode}}$ $259, $269, $439, $649

35. 1987, 2576, 3700, 4700, 5000, 7200, 8764, 9365

↑

median = $(4700+5000)\div 2=4850$

37. (a) $2(1199)+999+2(959)+1180+1168$
$+1029+949+1019=10,660$

$10,660\div 10=\$1066$

(b) $1066+0.825(1066)=\$1153.95$

39. $1153.95+377.79=\$1531.74$

41. (a) $1500+1700+1650+1300+1440$
$+1580+1820+1380+2900+6300=21,570$

$21,570\div 10=\$2157$

41. (b) 1300
1380
1440
1500
1580 ←
1650
1700
1820
2900
6300

median $= (1580 + 1650) \div 2 = \$1615$

(c) Median, because mean is affected by the high salary of \$6300.

Cumulative Review Problems

43. $\left.\frac{x}{2}+4\right|_{x=26} = \frac{26}{2}+4 = 13+4 = 17$

45. $\left.2x+1\right|_{x=5} = 2(5)+1 = 10+1 = 11$

47. $\frac{3}{8}\cdot 9200 = \$3450$

9.3 Exercises

1. From origin, move 2 right and 1 down.

3. (1994,4700), (1995,4800), (1996,4950), (1997,4850)

5. Lena, (4,60); Janie, (7,62); Mark, (9,66)

7. $(2,2)$ **9.** $(-1,4)$ **11.** $(3,-2)$ **13.** $(-2,-3)$

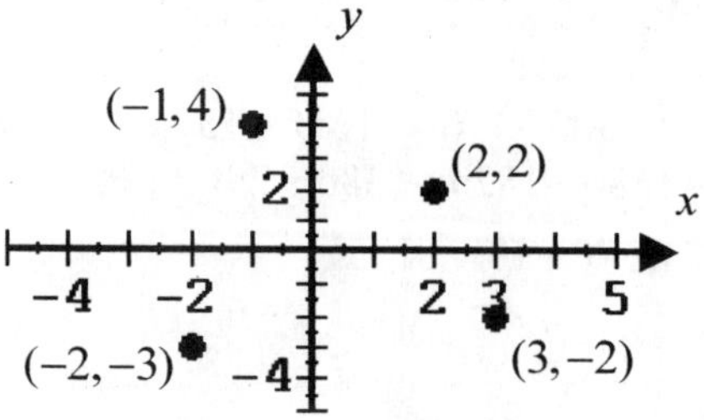

15. $(-1,2)$ **17.** $(5,-1)$ **19.** $(0,-2)$ **21.** $(5,0)$

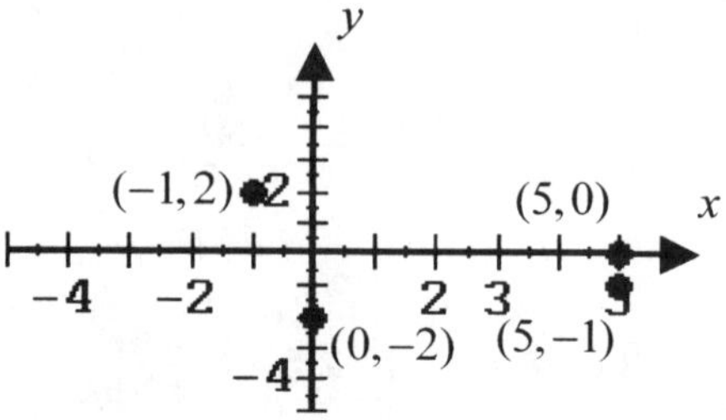

23. $\left(4\frac{1}{2},3\right)$ **25.** $\left(-1\frac{1}{2},-4\right)$ **27.** $\left(2\frac{1}{2},-1\right)$

29. $\left(-3\frac{1}{2},2\right)$

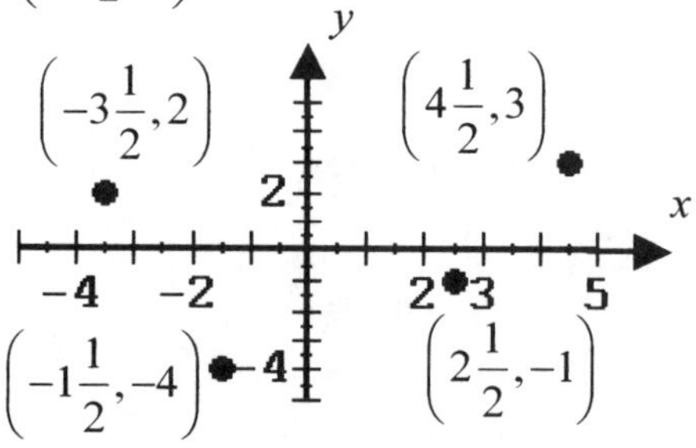

31. $K(-3,-3)$ **33.** $M(3,3)$

35. $O(-1,4)$ **37.** $Q\left(-3\frac{1}{2},4\right)$

39. $S\left(1\frac{1}{2},-3\right)$ **41.** $U(5,-1)$

43. (a) $(-4,-2)$

(b)

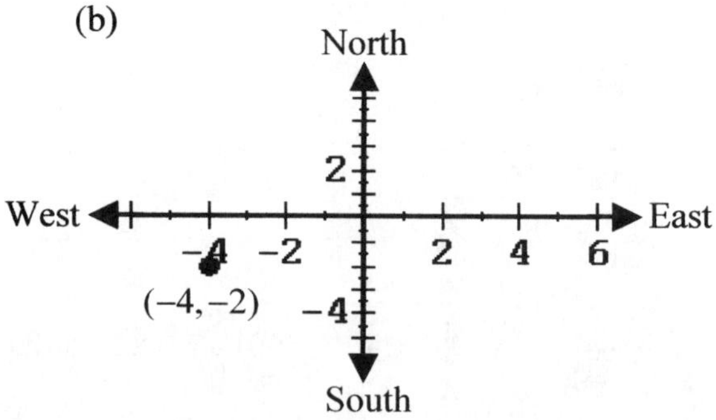

45. (a) (2,1)

(b)

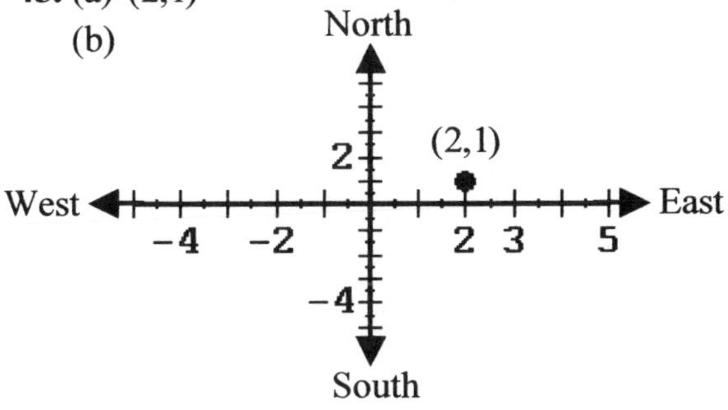

47. (2,4), (2,−1), (2,−3), (2,0)

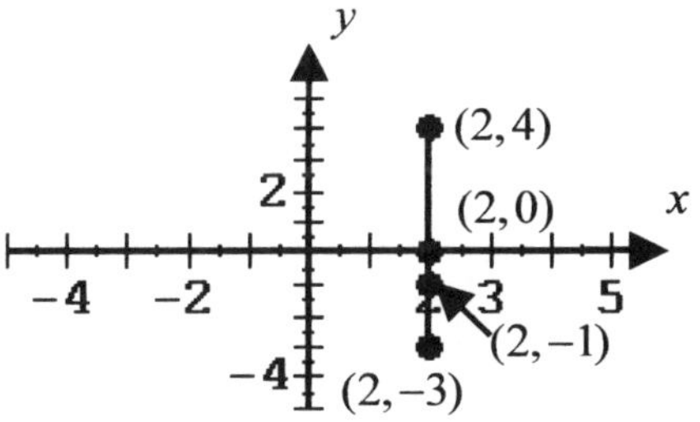

49. (1,3), (−5,3), (0,3), (−2,3)

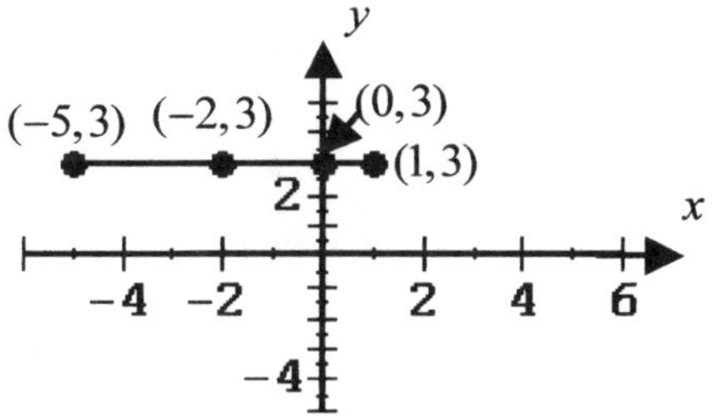

51. (0,−1), (−2,−1), (−4,−1), (4,−1)

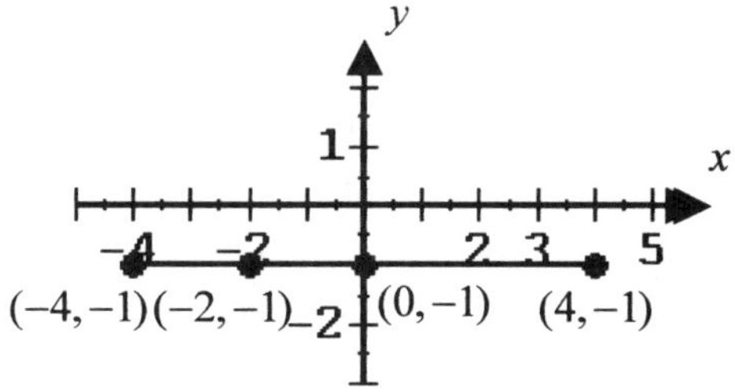

53. The figure is a rectangle.

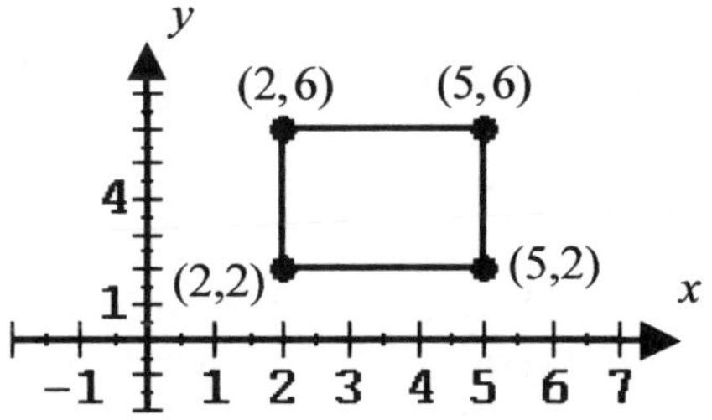

Cumulative Review Problems

55. $4x-3\big|_{x=2}=4(2)-3=8-3=5$

57. $2x-6\big|_{x=1}=2(1)-6=-4$

59. 39(150 million km) = 5850 million km

Putting Your Skills to Work, Problems for Individual Investigation

1. (1,3), (2,6), (4,12), (6,18)

2.

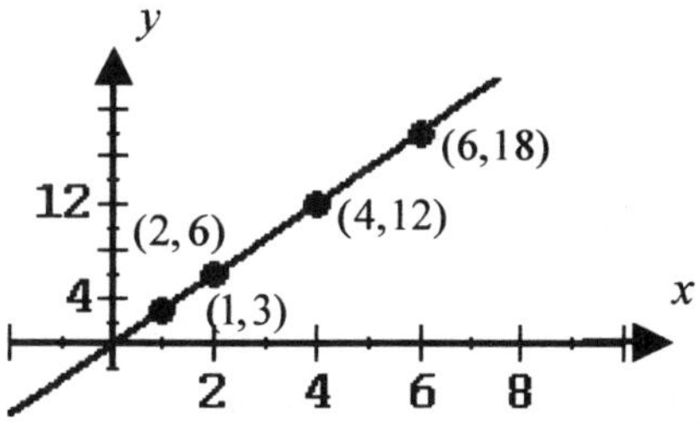

3. Positive correlation **4.** \$24

5. $P=3x$ where x is the number of items

6. $P=3(20)=\$60$

7. (1,3), (2,4), (3,5), (4,6), (5,8)

8.

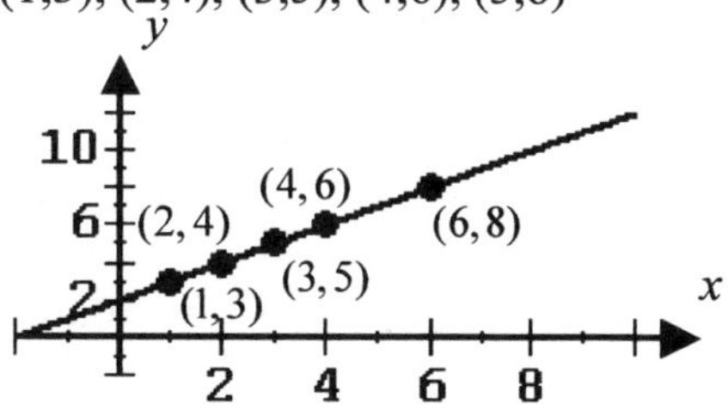

9. \$10

10. $C = x + 2$ ($x =$ number of components)

11. $C = 50 + 2 = \$52$

Putting Your Skills to Work, Problems for Group Investigation and Cooperative Study

1. (1,3), (2,6), (3,9), (4,12)

3. (1,2), (2,4), (3,6), (4,8)

2. 4.

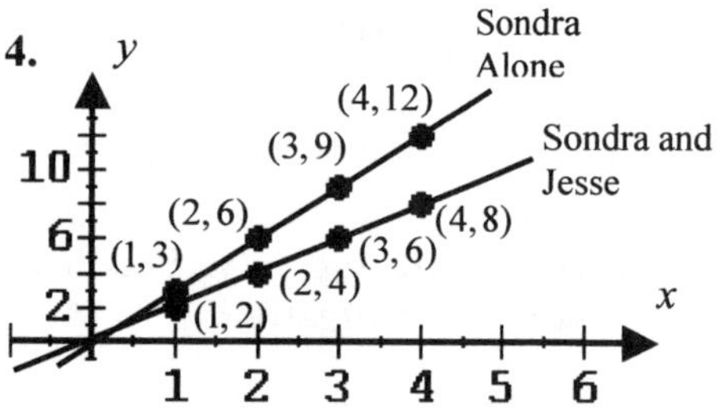

5. Sondra: $y = 3x$; together: $y = 2x$

6. (a) $y = 12(3) = 36$ hr, Sondra alone
(b) $y = 12(2) = 24$ hr, Together

7. $36(20) = \$720$ for Sondra alone
$24(20) + 24(15) = \$840$ together
It would cost less for Sondra alone.

8. Negative values make no sense here.

9.4 Exercises

1. $x + y = 4 \Rightarrow y = 4 - x$
$y = 4 - (-1) = 5 \Rightarrow (-1,5)$
$y = 4 - 0 = 4 \Rightarrow (0,4)$
$y = 4 - 1 = 3 \Rightarrow (1,3)$
$y = 4 - 4 = 0 \Rightarrow (4,0)$
$(-1,5)$, $(0,4)$, $(1,3)$, $(4,0)$ are solutions.

3. $x + y = 12 \Rightarrow y = 12 - x$
$y = 12 - (-1) = 13 \Rightarrow (-1,13)$
$y = 12 - 0 = 12 \Rightarrow (0,12)$
$y = 12 - 1 = 11 \Rightarrow (1,11)$
$y = 12 - 12 = 0 \Rightarrow (12,0)$
$(-1,13)$, $(0,12)$, $(1,11)$, $(0,12)$ are solutions to $x + y = 12$.

5. $y = 35x$
(a) $140 = 35x \Rightarrow x = \dfrac{140}{35} = 4$, $(4,140)$
(b) $280 = 35x \Rightarrow x = \dfrac{280}{35} = 8$, $(8,280)$

7. $y = 80x$
(a) $240 = 80x \Rightarrow x = 3 \Rightarrow (3,240)$
(b) $400 = 80x \Rightarrow x = 5 \Rightarrow (5,400)$

9. $x + 2y = 16 \Rightarrow y = \dfrac{16 - x}{2}$
$y = \left.\dfrac{16 - x}{2}\right|_{x=0} = 8 \Rightarrow (0,8)$
$x + 2(0) = 16 \Rightarrow x = 16 \Rightarrow (16,0)$
$x + 2(4) = 16 \Rightarrow x = 8 \Rightarrow (8,4)$

11. $x + y = 5$
$x + 2 = 5 \Rightarrow x = 3 \Rightarrow (3,2)$
$0 + y = 5 \Rightarrow y = 5 \Rightarrow (0,5)$
$1 + y = 5 \Rightarrow y = 4 \Rightarrow (1,4)$

13. $y = x + 2$
$y = 1 + 2 = 3 \Rightarrow (1,3)$
$3 = x + 2 \Rightarrow x = 1 \Rightarrow (1,3)$
$0 = x + 2 \Rightarrow x = -2 \Rightarrow (-2,0)$

15. $y = 5x + 3$

$y = 5(0) + 3 = 3 \Rightarrow (0, 3)$

$y = 5(-1) + 3 = -2 \Rightarrow (-1, -2)$

$y = 5(1) + 3 = 8 \Rightarrow (1, 8)$

17. $y = 5x - 3$

$y = 5(0) - 3 = -3 \Rightarrow (0, -3)$

$y = 5\left(\frac{3}{5}\right) - 3 = 0 \Rightarrow \left(\frac{3}{5}, 0\right)$

$y = 5(1) - 3 = 2 \Rightarrow (1, 2)$

19. $y = x + 6$

$y = 0 + 6 = 6 \Rightarrow (0, 6)$

$y = -6 + 6 = 0 \Rightarrow (-6, 0)$

$y = 1 + 6 = 7 \Rightarrow (1, 7)$

21. $y = 2x + 2$

$y = 2(-2) + 2 = -2 \Rightarrow (-2, -2)$

$y = 2(-1) + 2 = 0 \Rightarrow (-1, 0)$

$y = 2(0) + 2 = 2 \Rightarrow (0, 2)$

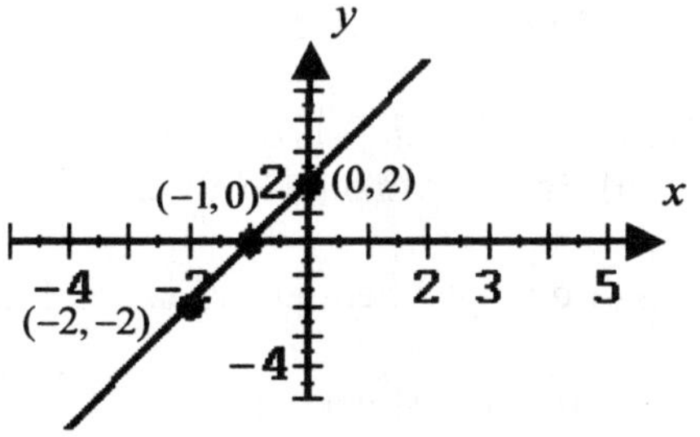

23. $y = -3x + 1$, $y = -3(-1) + 1 = 4 \Rightarrow (-1, 4)$

$y = -3(0) + 1 = 1 \Rightarrow (0, 1)$

$y = -3(1) + 1 = -2 \Rightarrow (1, -2)$

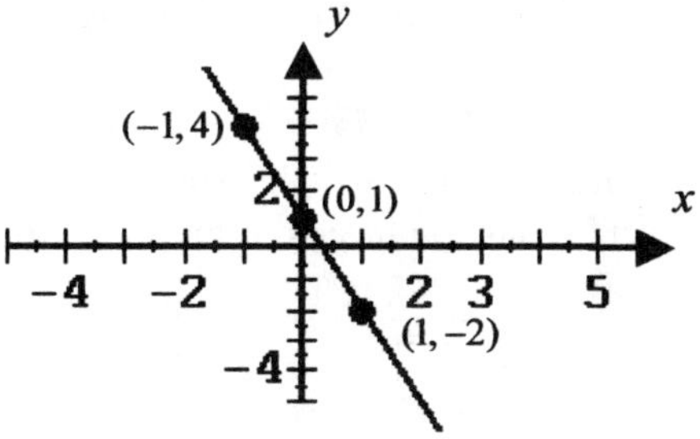

25. $y = 5x - 4$

$y = 5(0) - 4 = -4 \Rightarrow (0, -4)$

$y = 5(1) - 4 = 1 \Rightarrow (1, 1)$

$y = 5\left(\frac{2}{5}\right) - 4 = -2 \Rightarrow \left(\frac{2}{5}, -2\right)$

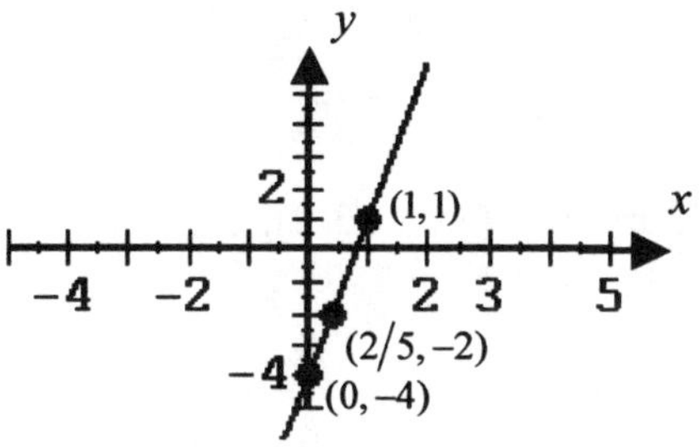

27. $y = 3x - 2$

$y = 3(-1) - 2 = -5 \Rightarrow (-1, -5)$

$y = 3(0) - 2 = -2 \Rightarrow (0, -2)$

$y = 3(1) - 2 = 1 \rightarrow\Rightarrow (1, 1)$

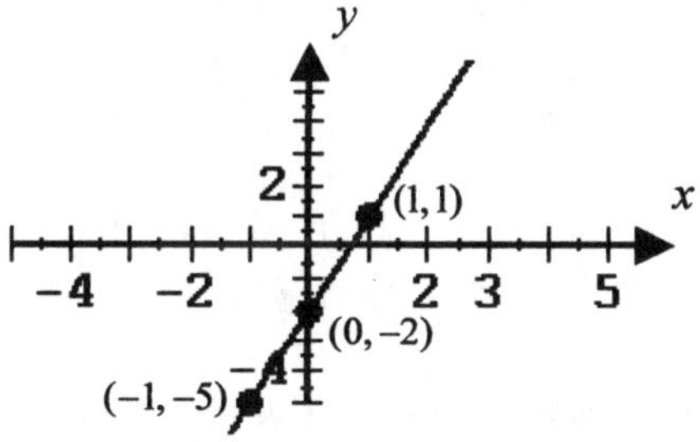

29. $y = -5x - 7$

$y = -5(-2) - 7 = 3 \Rightarrow (-2, 3)$

$y = -5\left(-\frac{3}{2}\right) - 7 = \frac{1}{2} \Rightarrow \left(-\frac{3}{2}, \frac{1}{2}\right)$

$y = -5(-1) - 7 = -2 \Rightarrow (-1, -2)$

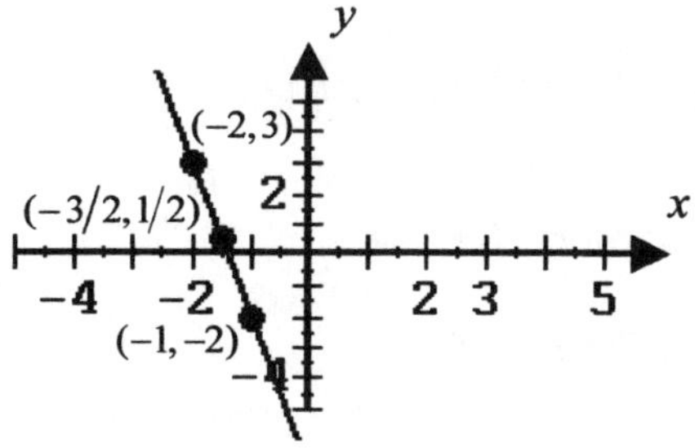

31. $y = 3 \Rightarrow (-4, 3),\ (0, 3),\ (5, 3)$

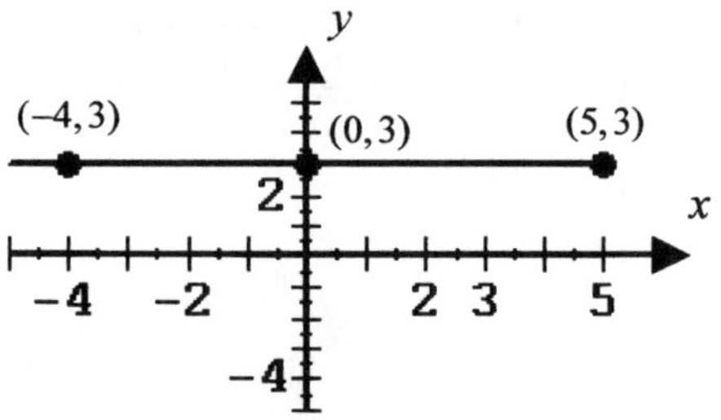

33. $x = 3 \Rightarrow (3, 5),\ (3, 0),\ (3, -2)$

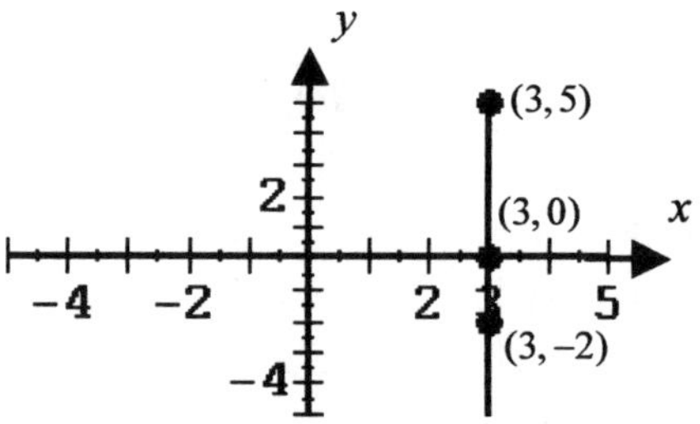

35. $(6, 0)$ is not a solution because it is not on the line formed by the other three points.

37. $(2, 4) = (2, 2 \cdot 2),\ (3, 6) = (3, 2 \cdot 3)$
$(4, 8) = (4, 2 \cdot 4) \Rightarrow y = 2x$

Cumulative Review Problems

39. $2x + 3 = 10 \Rightarrow 2x = 7 \Rightarrow x = \frac{7}{2}$

41. $-3x - 5 = 14 \Rightarrow -3x = 19 \Rightarrow x = -\frac{19}{3}$

43. $\frac{\$8}{\text{hr}} \cdot 25\frac{1}{2}\ \text{hr} = \$204$

Chapter 9 Review

1. $7(50) = 350$

2. $6(50) = 300$

3. $9(50) - 300 = 150$

4. $350 + 450 = 800$

5. 15%

6. 4%

7. $27\% + 30\% = 57\%$

8. $15\% + 8\% + 4\% = 27\%$

9. $0.08(2400) = \$192$

10. $0.15(2400) = \$360$

11. $(0.30 + 0.04)(2400) = \$816$

12. $(0.15 + 0.27)(2400) = \$1008$

13. (a) China with 1.2 billion
(b) India with 1.6 billion

14. (a) Brazil with 0.170 billion
(b) Brazil with 0.244 billion

15. Third largest behind China and India.

16. Fourth largest behind China, India, and the United States.

17. $1.3 - 1.2 = 0.1$ bil more people

18. $0.403 - 0.275 = 0.128$ bil more people

19. China: 0.1; India: 0.6; Indonesia: 0.1; United States: 0.128; Brazil: 0.074
India has the largest increase with 0.6 bil.

20. 0.1 bil

21. (a) \$1000 (b) \$600

22. (a) \$700 (b) \$100

23.

Item	Loss
Air	400
Alum. Wheels	250
Auto. Trans.	350
CD Player	1000
Cruise Control	200
Leather Trim	750
PL/PW	600
Power Seats	100
Power Sun Roof	550

CD Player, leather trim, power locks/window lost the most value.

24. Cruise control and power seats

25. $1000+400+800+1150+300$
$+1000+700+250+1000=6600$

$$\frac{6600}{9}=\$733.3\overline{3}$$

26. $600+150+450+150+100+250$
$+100+150+450=2400$

$$\frac{2400}{9}=\$266.6\overline{6}$$

27. 250
300
400
700
800 ← median = 800
1000
1000
1000
1150

28. 100
100
150
150
150 ← median = 150
250
450
450
600

29.

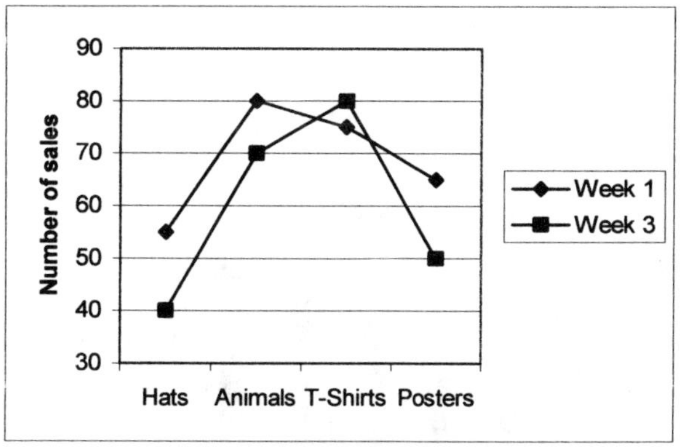

30.

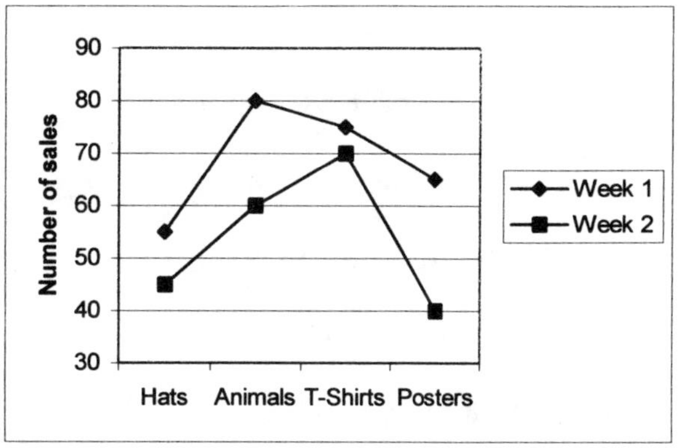

31.

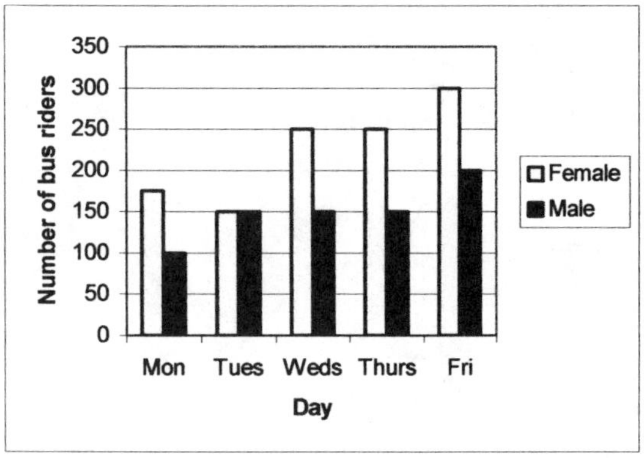

32.

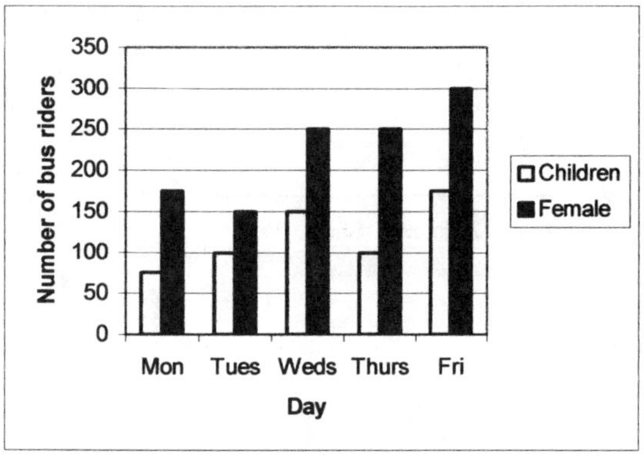

33. 57, 65, 69, 77, 82, 87, 88, 93, 100

median = 82

34. 58, 77, 79, 81, 83, 87, 88, 91, 104

median = 83

35. 0,1,4,5,9,18,19,19,20,21,22,25,27,36,38,43

$$\text{median} = \frac{19+20}{2} = 19.5$$

36. 0,3,9,13,14,15,16,18,19,21,24,25,26,28,31,36

$$\text{median} = \frac{18+19}{2} = 18.5$$

37. $\frac{86+83+88+95+97+100+81}{7} = 90°\text{F}$

38. $\frac{87+105+89+120+139+160+98}{7}$

$= \$114$

39. $\frac{76+20+91+57+42+21+75+82}{8} = 58$

40. $151+140+148+156+183+201$
$+205+228+231+237 = 1880$

$\frac{1880}{10} = 188$

41. $\underbrace{13,13,}_{\text{mode}}14,18,19,22$

42. $14,14,\underbrace{18,18,18,}_{\text{mode}}28,29$

43. (1994,5000), (1995,6000), (1996,5500), (1997,6250)

44. (1998,7000), (1999,6500), (2000,7500), (2001,7250)

45. 46. 47. 48. See graph below.

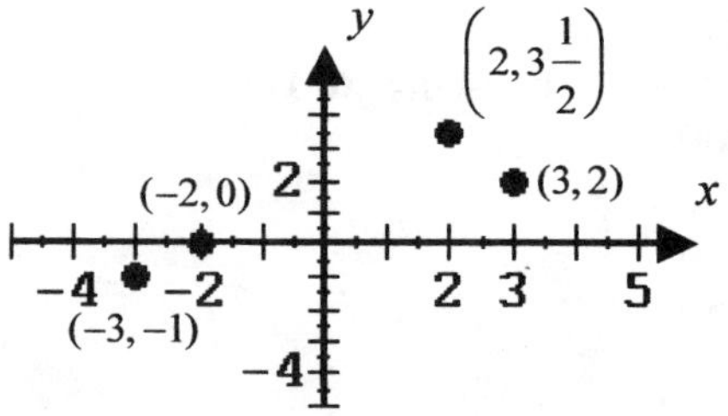

49. $R(-5,3)$

50. $S(-1,-1)$

51. $T\left(0,2\frac{1}{2}\right)$

52. $U(4,1)$

53. $(2,0),\ (2,3),\ (2,-1)$

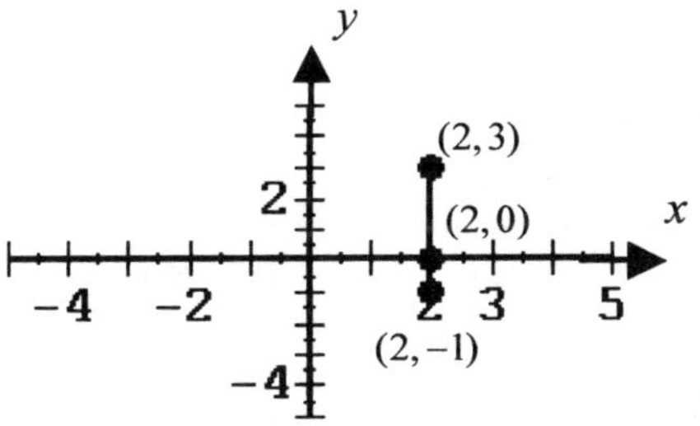

54. $(3,1),\ (-4,1),\ (0,1)$

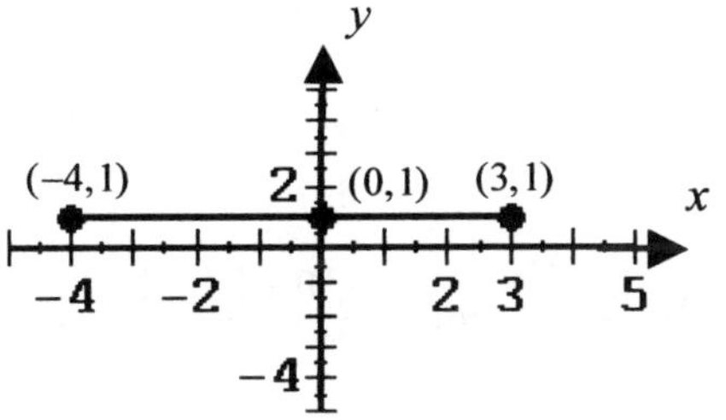

55. $x+y=8 \Rightarrow y=8-x$

$y=8-6=2 \Rightarrow (6,2)$

$y=8-8=0 \Rightarrow (8,0)$

$y=8-7=1 \Rightarrow (7,1)$

56. $x+y=3 \Rightarrow y=3-x$

$y=3-0=3 \Rightarrow (0,3)$

$y=3-2=1 \Rightarrow (2,1)$

$y=3-3=0 \Rightarrow (3,0)$

57. (a) $y=70x \Rightarrow 280=70x \Rightarrow x=4$

$(4,280)$

(b) $y=70x \Rightarrow 350=70x \Rightarrow x=5$

$(5,350)$

58. $y=2x-6 \Rightarrow x=\dfrac{y+6}{2}$

$x=\dfrac{-4+6}{2}=1 \Rightarrow (1,-4)$

$x=\dfrac{-6+6}{2}=0 \Rightarrow (0,-6)$

$x=\dfrac{-8+6}{2}=-1 \Rightarrow (-1,-8)$

59. $y=-6x+2 \Rightarrow x=\dfrac{2-y}{6}$

$x=\dfrac{2-2}{6}=0 \Rightarrow (0,2)$

$x=\dfrac{2-8}{6}=-1 \Rightarrow (-1,8)$

$x=\dfrac{2-(-4)}{6}=1 \Rightarrow (1,-4)$

60. $y=3x-1\big|_{x=-1}=-4 \Rightarrow (-1,-4)$

$y=3x-1\big|_{x=0}=-1 \Rightarrow (0,-1)$

$y=3x-1\big|_{x=1}=2 \Rightarrow (1,2)$

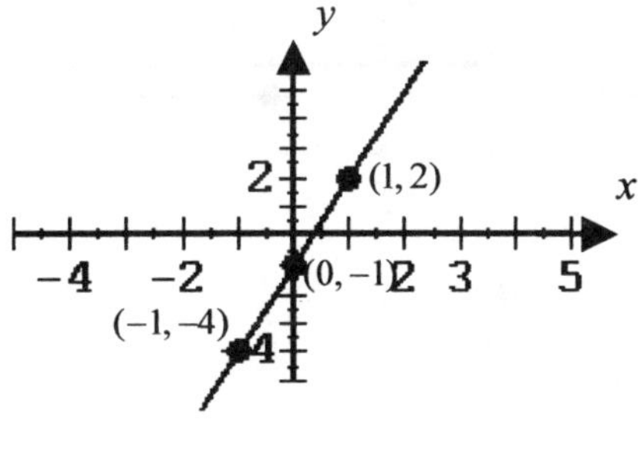

61. $y=-5x-4\big|_{x=-\frac{3}{2}}=-\dfrac{7}{2} \Rightarrow \left(-\dfrac{3}{2},-\dfrac{7}{2}\right)$

$y=-5x-4\big|_{x=-1}=1 \Rightarrow (-1,1)$

$y=-5x-4\big|_{x=0}=-4 \Rightarrow (0,-4)$

61.

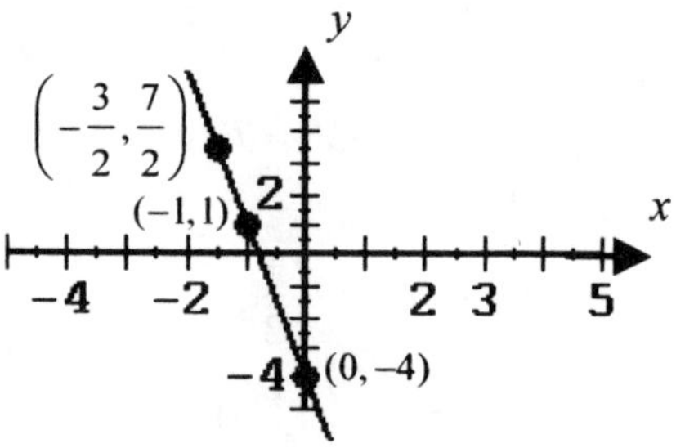

62. $y = 4x - 6\big|_{x=1} = -2 \Rightarrow (1, -2)$

$y = 4x - 6\big|_{x=\frac{3}{2}} = 0 \Rightarrow \left(\frac{3}{2}, 0\right)$

$y = 4x - 6\big|_{x=2} = 2 \Rightarrow (2, 2)$

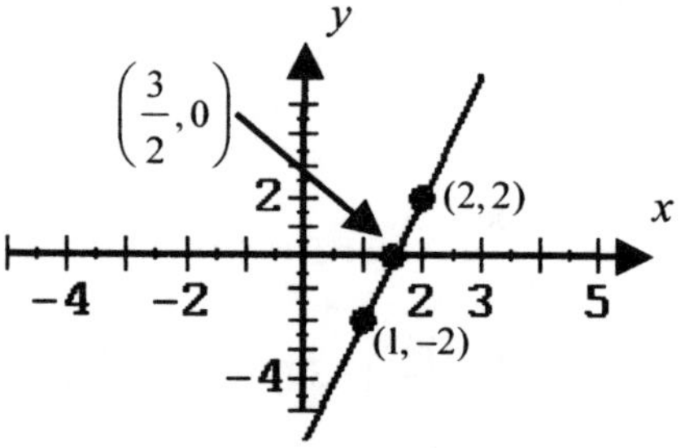

63. $y = -1 \Rightarrow (-5, -1),\ (2, -1),\ (5, -1)$

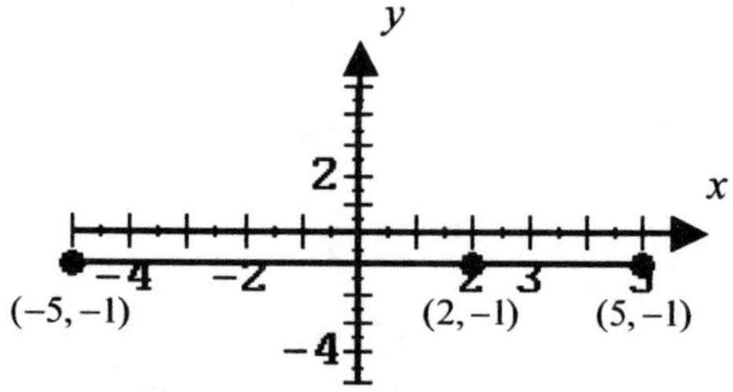

Chapter 9 Test

1. Age 18-20, 44%.

2. $44\% + 33\% = 77\%$

3. $44\% + 6\% = 50\%$

4. $0.07(5000) = 350$ students

5. 11; 12

6. 6; 9

7. Otsego, Lake, NY

8. Baikal Lake, Russia

9. Honda and Toyota do not have a road-side assistance plan.

10. 10 yr

11. 3 more yr

12. (a) Lexus (b) Volkswagen (c) Volkswagen (d) Lexus

13. Lexus: $\dfrac{4+6+6+4}{4} = 5$ yr

Honda: $\dfrac{3+3+5}{3} = 3.\bar{6}$ yr

14. Toyota: $\dfrac{3+5+5}{3} = 4.\bar{3}$ yr

Volkswagen: $\dfrac{2+10+12+2}{4} = 6.5$ yr

15. $\dfrac{89+76+85+91+83+90}{6} \approx 85.7$

16. 76, 83, 85, 89, 90, 91

$\uparrow$

median $= \dfrac{85+89}{2} = 87$

17. $(3, 5)$ See graph below.

18. $(0, 0)$ See graph below.

19. $(2, -1)$ See graph below.

20. $(-3,0)$ See graph below.

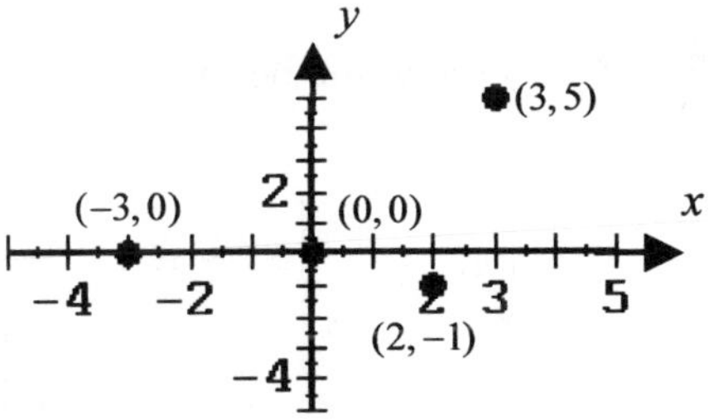

21. $A(4,2)$

22. $B(0,-2)$

23. $C(3,-1)$

24. $D(-1,-3)$

25. $(1,2)$, $(-3,2)$, $(0,2)$, $(4,2)$

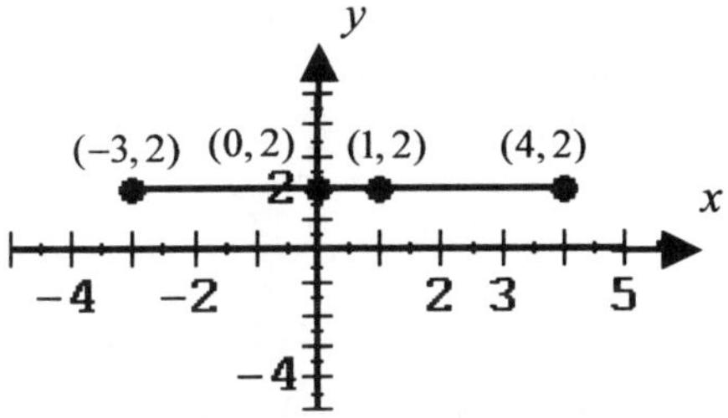

26. $y=4x-2\big|_{x=0}=-2\Rightarrow(0,-2)$

$y=4x-2\big|_{x=\frac{1}{2}}=0\Rightarrow\left(\frac{1}{2},0\right)$

$y=4x-2\big|_{x=1}=2\Rightarrow(1,2)$

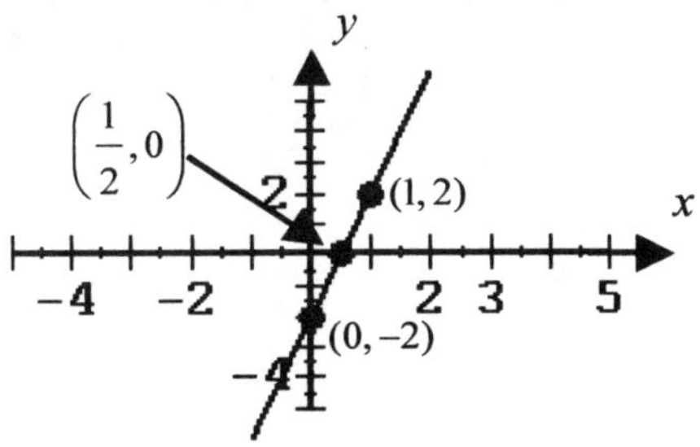

27. $y=3\Rightarrow(-3,3)$, $(0,3)$, $(3,3)$

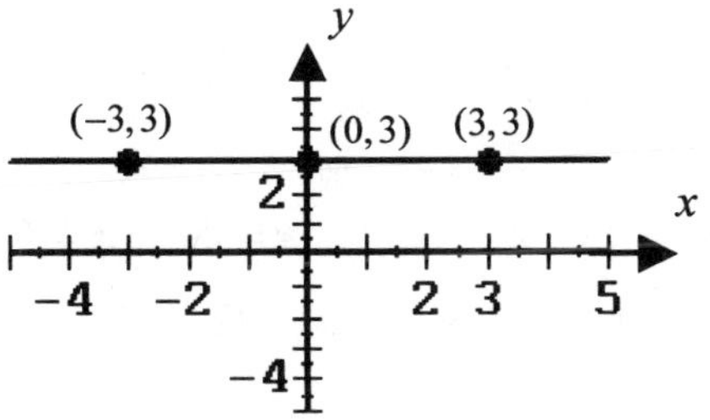

Cumulative Test for Chapters 1-9

1. $5\cdot5\cdot5\cdot5\cdot5\cdot5=5^6$

2. $4^3=4\cdot4\cdot4=64$

3. $3^2-2+7=9+5=14$

4. $20\div4+5(2+1)=5+5(3)=5+15=20$

5. $-2^3=-2\cdot2\cdot2=-8$

6. $98=2\cdot7\cdot7=2\cdot7^2$

7. $\frac{285\text{ mi}}{12\text{ gal}}\approx23.8\text{ mi/gal}$

8. $\frac{xy^2}{18}\cdot\frac{9}{x}=\frac{y^2}{2}$

9. The reciprocal of -11 is $\frac{1}{-11}=-\frac{1}{11}$

10. $\frac{2}{7}\div\frac{-8}{14}=-\frac{2}{7}\cdot\frac{14}{8}=-\frac{1}{2}$

11. $10x=120\Rightarrow x=12$

check: $10(12)\overset{?}{=}120$, $120=120$

12. $\frac{x}{3} = 12 \Rightarrow x = 3(12) = 36$

check: $\frac{36}{3} \stackrel{?}{=} 12,\ 12 = 12$

13. $-5x - 2 = -22 \Rightarrow 5x = 20 \Rightarrow x = 4$

check: $-5(4) - 2 \stackrel{?}{=} -22,\ -22 = -22$

14. $2(2x^3 - x + 6) = 4x^3 - 2x + 12$

15. $(3x^2 - 2x - 8) + (x^2 + 4x - 3)$
$= 3x^2 - 2x - 8 + x^2 + 4x - 3$
$= 4x^2 + 2x - 11$

16. $(3x + 1) - (5x - 2) = 3x + 1 - 5x + 2$
$= -2x + 3$

17. 20%

18. 10%

19. 60% − 5% = 55%

20. 45% − 15% = 30%

21. $\frac{76 + 83 + 72 + 86 + 88}{5} = 81$

22. 68, 80, 90, 105, 115, 121

median $= \frac{90 + 105}{2} = 97.5$

23. $1, 2, 6, 7, \underbrace{8, 8}_{\text{mode}}, 9$

24. (6,1) See graph below.

25. (2,3) See graph below.

26. (0,−2) See graph below.

27. (−4,−3) See graph below.

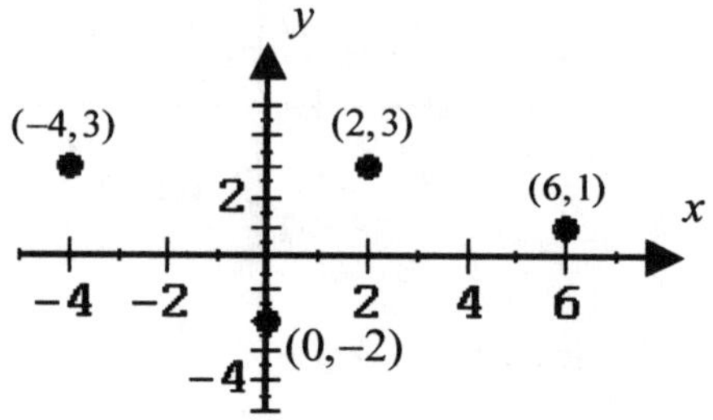

28. $y = x + 3\big|_{x=-1} = 2 \Rightarrow (-1, 2)$
$y = x + 3\big|_{x=0} = 3 \Rightarrow (0, 3)$
$y = x + 3\big|_{x=1} = 4 \Rightarrow (1, 4)$

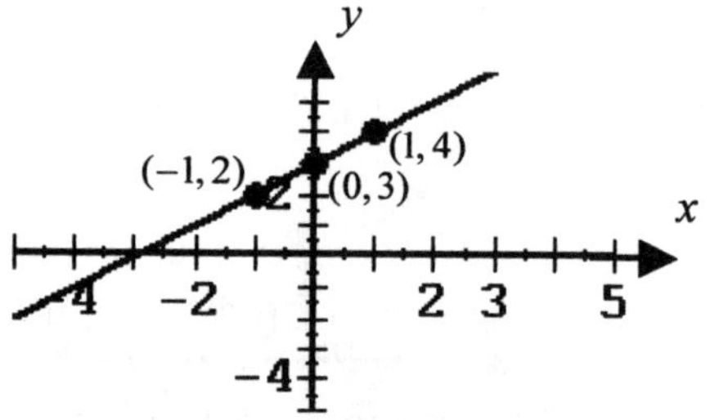

29. $y = -3x - 4\big|_{x=-2} = 2 \Rightarrow (-2, 2)$
$y = -3x - 4\big|_{x=-1} = -1 \Rightarrow (-1, -1)$
$y = -3x - 4\big|_{x=0} = -4 \Rightarrow (0, -4)$

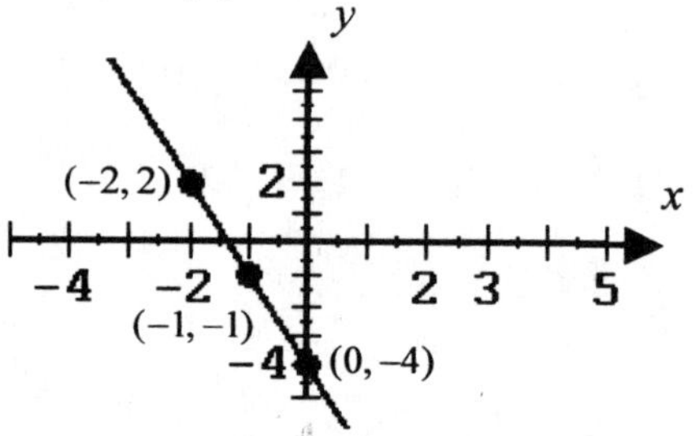

Chapter 10

Pretest Chapter 10

1. $370 \text{ min} = 360 \text{ min} \cdot \dfrac{\text{hr}}{60 \text{ min}} + 10 \text{ min}$
$= 6 \text{ hr } 10 \text{ min}$

2. $10\frac{1}{4} \text{ ft} \cdot \dfrac{12 \text{ in.}}{\text{ft}} = 123 \text{ in.}$

3. $5000 \text{ m} \cdot \dfrac{\text{km}}{1000 \text{ m}} = 5 \text{ km}$

4. $35 \text{ m} \cdot \dfrac{\text{ft}}{0.305 \text{ m}} \approx 114.8 \text{ ft}$

5. $7 \text{ mm} \cdot \dfrac{0.394 \text{ in.}}{\text{cm}} \cdot \dfrac{\text{cm}}{10 \text{ mm}} = 0.2758 \text{ in.}$

6. $\angle DBC, \angle CBD$

7. (a) $180° - 49° = 131°$
(b) $90° - 49° = 41°$

8. $61° = \angle c = \angle e$
$180° - 61° = 119° = \angle b = \angle d$

9. $n > 0,\ n^2 = 9 \Rightarrow n = 3$

10. $n > 0,\ n \cdot n = 4 \Rightarrow n = 2$

11. $\sqrt{36} = 6$

12. $\sqrt{100} + \sqrt{25} = 10 + 5 = 15$

13. $c = \sqrt{3^2 + 4^2} = \sqrt{16 + 9} = \sqrt{25} = 5$

14. $C = \pi d \approx 3.14(7) = 21.98 \text{ m}$

15. $A = \pi r^2 \approx 3.14(3)^2 = 28.26 \text{ in.}^2$

16. $A = \pi r^2 \approx 3.14(5.3)^2 \approx 88.2 \text{ cm}^2$

17. $A = \pi r^2 = \pi(2)^2 = 4\pi \text{ ft}^2 \cdot \dfrac{\text{yd}^2}{9 \text{ ft}^2}$

$4(3.14) \text{ ft}^2 \cdot \dfrac{\text{yd}^2}{9 \text{ ft}^2} \cdot \dfrac{\$15}{\text{yd}^2} = \$20.93$

18. $V = \dfrac{4\pi r^3}{3} \approx \dfrac{4(3.14)(6)^3}{3} = 904.32 \text{ in.}^3$

19. $V = \dfrac{Bh1}{3} \approx \dfrac{(4)(5)(3)}{3} = 20 \text{ m}^3$

20. $\dfrac{h}{30} = \dfrac{6}{12} \Rightarrow h = 15 \text{ ft}$

10.1 Exercises

1. 1 ft = 12 in.

3. 2 pints = 1 quart

5. 1 ton = 2000 lb

7. 4 qt = 1 gal

9. 7 days = 1 week

11. 60 sec = 1 min

13. $15 \text{ ft} \cdot \dfrac{\text{yd}}{3 \text{ ft}} = 5 \text{ yd}$

15. $10{,}560 \text{ ft} \cdot \dfrac{1 \text{ mi}}{5280 \text{ ft}} = 2 \text{ mi}$

17. $8 \text{ pints} \cdot \dfrac{1 \text{ qt}}{2 \text{ pints}} = 4 \text{ qt}$

19. $12 \text{ ft} \cdot \frac{12 \text{ in.}}{1 \text{ ft}} = 144 \text{ in.}$

21. $87 \text{ in.} \cdot \frac{1 \text{ ft}}{12 \text{ in}} = 7.25 \text{ ft}$

23. $13 \text{ tons} \cdot \frac{2000 \text{ lb}}{1 \text{ ton}} = 26{,}000 \text{ lb}$

25. $7 \text{ gal} \cdot \frac{4 \text{ qt}}{1 \text{ gal}} = 28 \text{ qt}$

27. $7 \text{ gal} \cdot \frac{4 \text{ qt}}{1 \text{ gal}} = 28 \text{ qt}$

29. $660 \text{ min} \cdot \frac{1 \text{ hr}}{60 \text{ min}} = 11 \text{ hr}$

31. $6 \text{ tons} \cdot \frac{2000 \text{ lb}}{1 \text{ ton}} = 12{,}000 \text{ lb}$

33. $26 \text{ oz} \cdot \frac{1 \text{ lb}}{16 \text{ oz}} \cdot \frac{\$6}{\text{lb}} = \$9.75$

35. $2.745 \text{ mi} \cdot \frac{5280 \text{ ft}}{1 \text{ mi}} \approx 14{,}490 \text{ ft}$

37. $44 \text{ cm} \cdot \frac{10 \text{ mm}}{1 \text{ cm}} = 440 \text{ mm}$

39. $7.6 \text{ km} \cdot \frac{1000 \text{ m}}{\text{km}} = 7600 \text{ m}$

41. $1.76 \text{ kL} \cdot \frac{1000 \text{ L}}{\text{kL}} \cdot \frac{1000 \text{ mL}}{1 \text{ L}}$
$= 1{,}760{,}000 \text{ mL}$

43. $5261 \text{ mL} \cdot \frac{1 \text{ L}}{1000 \text{ mL}} \cdot \frac{\text{kL}}{1000 \text{ L}}$
$= 0.005261 \text{ kL}$

45. $294 \text{ g} \cdot \frac{\text{kg}}{1000 \text{ g}} = 0.294 \text{ kg}$

47. $6328 \text{ mg} \cdot \frac{1 \text{ g}}{1000 \text{ mg}} = 6.328 \text{ g}$

49. $0.83 \text{ kg} \cdot \frac{1000 \text{ g}}{1 \text{ kg}} \cdot \frac{1000 \text{ mg}}{1 \text{ g}}$
$= 830{,}000 \text{ mg}$

51. $18 \text{ mL} = 0.018 \text{ L} = 0.000018 \text{ kL}$

53. $49 \text{ mg} = 0.049 \text{ g} = 0.000049 \text{ kg}$

55. $7183 \text{ mg} = 7.183 \text{ g} = 0.007183 \text{ kg}$

57. $83 \text{ mm} = 8.3 \text{ cm} = 0.083 \text{ m}$

59. $7812 \text{ mm} = 7.812 \text{ m} = 0.007812 \text{ km}$

61. $0.81 \text{ cm} = 0.0081 \text{ m} = 0.0000081 \text{ km}$

63. $6 \text{ kg} \cdot \frac{1000 \text{ g}}{1 \text{ kg}} \cdot \frac{\$95.50}{\text{g}} = \$573{,}000$

65. (a) $94{,}380{,}000 \text{ cm} \cdot \frac{1 \text{ m}}{100 \text{ cm}}$
$= 943{,}800 \text{ m}$

(b) $943{,}800 \text{ m} \cdot \frac{1 \text{ km}}{1000 \text{ m}} = 943.8 \text{ km}$

67. (a) $335 \text{ m} \cdot \frac{1 \text{ km}}{1000 \text{ m}} = 0.335 \text{ km}$

(b) $335 \text{ m} \cdot \frac{100 \text{ cm}}{1 \text{ m}} = 33{,}500 \text{ cm}$

69. $3 \text{ yr} \cdot \frac{365 \text{ day}}{1 \text{ yr}} \cdot \frac{24 \text{ hr}}{1 \text{ day}} = 26{,}280 \text{ hr}$

71. $528 \text{ megabytes} = 528{,}000{,}000 \text{ bytes}$

73. 24.9 gigabytes = 24,900,000,000 bytes

Cumulative Review Problems

75. $a = 0.23(250) = 57.5$

77. $0.08(8960) = \$716.80$

10.2 Exercises

1. $7 \text{ ft} \cdot \frac{0.305 \text{ m}}{\text{ft}} \approx 2.14 \text{ m}$

3. $9 \text{ in.} \cdot \frac{2.54 \text{ cm}}{\text{in.}} = 22.86 \text{ cm}$

5. $14 \text{ m} \cdot \frac{1.09\text{yd}}{1 \text{ m}} = 15.26 \text{ yd}$

7. $26.5 \text{ m} \cdot \frac{1.09\text{yd}}{1 \text{ m}} \approx 26.89 \text{ yd}$

9. $15 \text{ km} \cdot \frac{0.62 \text{ mi}}{1 \text{ km}} = 9.3 \text{ mi}$

11. $24 \text{ yd} \cdot \frac{0.914 \text{ m}}{\text{yd}} \approx 21.94 \text{ m}$

13. $82 \text{ mi} \cdot \frac{1.61 \text{ km}}{1 \text{ mi}} = 132.02 \text{ km}$

15. $25 \text{ m} \cdot \frac{3.28 \text{ ft}}{\text{m}} = 82 \text{ ft}$

17. $17.5 \text{ cm} \cdot \frac{0.394 \text{ in.}}{\text{cm}} \approx 6.90 \text{ in.}$

19. $200 \text{ m} \cdot \frac{1.09 \text{ yd}}{1 \text{ m}} = 218 \text{ yd}$

21. $5 \text{ gal} \cdot \frac{3.79 \text{ L}}{1 \text{ gal}} = 18.95 \text{ L}$

23. $19 \text{ L} \cdot \frac{0.264 \text{ gal}}{1 \text{ L}} \approx 5.02 \text{ gal}$

25. $4.5 \text{ L} \cdot \frac{1.06 \text{ qt}}{1 \text{ L}} = 4.77 \text{ qt}$

27. $32 \text{ lb} \cdot \frac{0.454 \text{ kg}}{1 \text{ lb}} \approx 14.53 \text{ kg}$

29. $7 \text{ oz} \cdot \frac{28.35 \text{ g}}{\text{oz}} = 198.45 \text{ g}$

31. $16 \text{ kg} \cdot \frac{2.2 \text{ lb}}{\text{kg}} = 35.2 \text{ lb}$

33. $126 \text{ g} \cdot \frac{0.0353 \text{ oz}}{\text{g}} \approx 4.45 \text{ oz}$

35. $166 \text{ cm} \cdot \frac{1 \text{ in.}}{2.54 \text{ cm}} \cdot \frac{1 \text{ ft}}{12 \text{ in.}} = 5.45 \text{ ft}$

37. $16.5 \text{ ft} \cdot \frac{12 \text{ in}}{\text{ft}} \cdot \frac{2.54 \text{ cm}}{\text{in.}} = 502.92 \text{ cm}$

39. $50 \frac{\text{km}}{\text{hr}} \cdot \frac{0.62 \text{ mi}}{\text{km}} = 31 \frac{\text{mi}}{\text{hr}}$

41. $60 \frac{\text{mi}}{\text{hr}} \cdot \frac{1.61 \text{ km}}{\text{mi}} = 96.6 \frac{\text{km}}{\text{hr}}$

43. $13 \text{ mm} \cdot \frac{\text{cm}}{10 \text{ mm}} \cdot \frac{1 \text{ in.}}{2.54 \text{ cm}} \approx 0.51 \text{ in.}$

45. $F = 1.8C + 32 = 1.8(40) + 32$
$= 104\ ^\circ\text{F}$

47. $F = 1.8C + 32 = 1.8(85) + 32$
$= 185\ °\text{F}$

49. $F = 1.8C + 32 = 1.8(12) + 32$
$= 53.6\ °\text{F}$

51. $C = \dfrac{5F - 160}{9} = \dfrac{5(68) - 160}{9} = 20\ °\text{C}$

53. $C = \dfrac{5F - 160}{9} = \dfrac{5(168) - 160}{9} \approx 75.56\ °\text{C}$

55. $C = \dfrac{5F - 160}{9} = \dfrac{5(86) - 160}{9} = 30\ °\text{C}$

57. $67 \text{ mi} \cdot \dfrac{1.61 \text{ km}}{1 \text{ mi}} + 36 \text{ km} = 143.87 \text{ km}$

59. $15 \text{ gal} \cdot \dfrac{3.79 \text{ L}}{1 \text{ gal}} - 38 \text{ L} = 18.85 \text{ L}$

61. $635 \text{ kg} \cdot \dfrac{2.2 \text{ lb}}{1 \text{ kg}} = 1397 \text{ lb}$

63. $F = 1.8C + 32 = 1.8(19) + 32$
$= 66.2\ °\text{F}$
$F = 1.8C + 32 = 1.8(45) + 32$
$= 113\ °\text{F}$

65. $C = \dfrac{5F - 160}{9} = \dfrac{5(6188) - 160}{9}$
$= 3420\ °\text{C}$

67. $0.768 \text{ oz} \cdot \dfrac{28.35 \text{ g}}{1 \text{ oz}} \approx 21.773 \text{ g}$

Cumulative Review Problems

69. $2^3 \times 6 - 4 + 3 = 8 \times 6 - 1 = 48 - 1 = 47$

71. $2^2 + 3^2 + 4^3 + 2 \times 7 = 4 + 9 + 64 + 14 = 91$

Putting Your Skills to Work, Problems for Individual Investigation

1. $880 \text{ km} \cdot \dfrac{0.62 \text{ mi}}{1 \text{ km}} = 545.6 \text{ mi}$

2. $16 \text{ gal} \cdot \dfrac{3.79 \text{ L}}{1 \text{ gal}} \approx 60.6 \text{ L}$

3. Meat: $5 \text{ lb} \cdot \dfrac{0.454 \text{ kg}}{1 \text{ lb}} = 2.27 \text{ kg}$

Carrots: $2 \text{ lb} \cdot \dfrac{0.454 \text{ kg}}{1 \text{ lb}} \approx 0.91 \text{ kg}$

Apples: $3 \text{ lb} \cdot \dfrac{0.454 \text{ kg}}{1 \text{ lb}} \approx 1.36 \text{ kg}$

Potatoes: $10 \text{ lb} \cdot \dfrac{0.454 \text{ kg}}{1 \text{ lb}} = 4.54 \text{ kg}$

4. $F = 1.8C + 32 = 1.8(25) + 32$
$= 77\ °\text{F}$

5. $\$70 \cdot \dfrac{1.75 \text{ marks}}{\$} = 122.5 \text{ marks}$

Putting Your Skills to Work, Problems for Group Investigation and Cooperative Study

1. $30\ \dfrac{\text{mi}}{\text{gal}} \cdot \dfrac{1.61 \text{ km}}{\text{mi}} \cdot \dfrac{0.264 \text{ gal}}{\text{L}} \approx 12.75\ \dfrac{\text{km}}{\text{L}}$

2. Meat: $15\ \dfrac{\text{marks}}{\text{kg}} \cdot \dfrac{0.454 \text{ kg}}{1 \text{ lb}} \cdot \dfrac{\$}{1.75 \text{ marks}}$
$= \$3.89/\text{lb}$

Apples:
$3.5\ \dfrac{\text{marks}}{\text{kg}} \cdot \dfrac{0.454 \text{ kg}}{1 \text{ lb}} \cdot \dfrac{\$}{1.75 \text{ marks}}$
$= \$0.91/\text{lb}$

3. Potatoes:

$$\frac{12\text{marks}}{10\text{ kg}}\cdot\frac{0.454\text{ kg}}{1\text{ lb}}\cdot\frac{\$}{1.75\text{ marks}}$$
$$=\$0.31/\text{lb}$$

10.3 Exercises

1. Vertex: The point at which the two sides of an angle meet.

3. Obtuse angle: An angle whose measure is between 90° and 180°.

5. Adjacent angles: Two angles formed by intersecting lines that share a common side.

7. Line u is called a transversal since it intersects two or more lines at different points.

9. $\angle ABC$ is called a straight angle.

11. $\angle ABD$ is called an obtuse angle.

13. If the sum of two angles is 180° then these angles are called supplementary angles.

15. An angle whose measure is between 0° and 90° is called an acute angle.

17. $\angle a$ and $\angle c$ are called vertical angles.

19. $\angle d$ is equal to $\angle b$.

21. $\angle a+\angle b=180°$.

23. Name the angles that are adjacent to $\angle c$: $\angle d$ and $\angle$b.

25. $\angle a$ and $\angle d$ are called alternate interior angles.

27. $\angle b$ is equal to the following two angles: $\angle c$ and $\angle f$.

29. $\angle f$ is equal to the following two angles: $\angle b$ and $\angle c$.

31. $\angle e$ and $\angle d$ are called corresponding angles.

33. $\angle b$ and $\angle f$ are called vertical angles.

35. $\angle XYZ$, $\angle ZYX$, or $\angle Y$.

37. $\angle TSV$ or $\angle VST$

39.(a) supplement of $\angle M=180°-29°=151°$

(b) complement of $\angle M=90°-29°=61°$

41.(a) supplement of $\angle X=180°-45°=135°$

(b) complement of $\angle X=90°-45°=45°$

43. $x+(x+15°)=90°\Rightarrow 2x=90°-15°$

$2x=75°\Rightarrow x=37.5°$

45. $n+(n-5°)=90°\Rightarrow 2n=90°+5°$

$2n=95°\Rightarrow n=47.5°$

47. $2m+m=90°\Rightarrow 3m=90°\Rightarrow m=30°$

49. $\angle w$ and $\angle y$ are adjacent to $\angle x$.

51. $\angle w+\angle x=180°$.

53. $\angle x=\angle z\Rightarrow\angle z=65°, \angle w=180°-\angle x$

$\angle w=180°-65°=115°=\angle y$

55. $\angle w = 45° = \angle y,\ \angle x = 180° - \angle w$
$\angle x = 180° - 45° = 135° = \angle z$

57. $\angle p = \angle o = 43°,\ \angle o = \angle n = 43°$
$\angle n = \angle l = 43°,\ \angle m = 180° - 43° = 137°$

59. $\angle b = 180° - 45° = 135°,\ \angle a = \angle c = 45°$
$\angle c = \angle e = 45°,\ \angle d = 180° - 45° = 135°$

61. $\angle y = \angle v = \angle u = \angle r = 99°$
$\angle x = \angle w = \angle t = \angle s = 180° - 99° = 81°$

63. $\angle z = 180° - \angle w = 180° - 52° = 128°$
$\angle w = \angle x = 52°$
$\angle y = 90° - \angle x = 90° - 52° = 38°$

Cumulative Review Problems

65. $V = LWH = 10(4)(8) = 320\text{ m}^3$

67. $70\text{ sq}\cdot\dfrac{\text{in.}}{14\text{ sq}} + 2(2)\text{ in.} = 9\text{ in.}$

$84\text{ sq}\cdot\dfrac{\text{in.}}{14\text{ sq}} + 2(2)\text{ in.} = 10\text{ in.}$

9 in. by 10 in.

10.4 Exercises

1. (a) $\sqrt{81}$: What is the square root of 81?
(b) $n\cdot n = 81$: What number multiplied by itself equals 81?
(c) $n^2 = 81$: What number squared equals 81?

3. $6^2 = 36,\ 7^2 = 49 \Rightarrow 44$ is not a perfect square.

5. $6^2 = 36 \Rightarrow 36$ is a perfect square.

7. $9^2 = 81 \Rightarrow 81$ is a perfect square.

9. $\left(\dfrac{1}{2}\right)^2 = \dfrac{1}{4},\ \left(\dfrac{1}{3}\right)^2 = \dfrac{1}{9} \Rightarrow \dfrac{1}{5}$ is not a perfect square.

11. (a) $n^2 = 64 \Leftrightarrow n = \sqrt{64}$
(b) $\sqrt{64}$ is the positive square root of 64.
(c) $n\cdot n = 64 \Leftrightarrow n = \sqrt{64}$

13. $\sqrt{49} = 7$

15. $\sqrt{64} = 8$

17. $\sqrt{144} = 12$

19. $\sqrt{225} - \sqrt{16} = 15 - 4 = 11$

21. $\sqrt{9} + \sqrt{49} = 3 + 7 = 10$

23. $\sqrt{\dfrac{25}{49}} = \dfrac{5}{7}$

25. $\sqrt{\dfrac{9}{81}} = \dfrac{3}{9} = \dfrac{1}{3}$

27. $\sqrt{\dfrac{36}{49}} = \dfrac{6}{7}$

29. $A = s^2 = 121 \Rightarrow s = 11\text{ ft}$

31. $A = s^2 = 144 \Rightarrow s = 12\text{ ft}$

33. $\sqrt{31} \approx 5.568$

35. $\sqrt{69} \approx 8.307$

37. $\sqrt{80} \approx 8.944$

39. $\sqrt{90} \approx 9.487$

41. $c^2 = 3^2 + 4^2 = 9 + 16 = 25 \Rightarrow c = 5$ in.

43. $c^2 = 8^2 + 3^2 = 64 + 9 = 73$
$c = \sqrt{73} \approx 8.544$ yd

45. $16^2 = a^2 + 5^2 \Rightarrow a^2 = 256 - 25 = 231$
$a = \sqrt{231} \approx 15.199$ ft

47. $13^2 = 8^2 + b^2 \Rightarrow b^2 = 169 - 64 = 105$
$b = \sqrt{105} \approx 10.247$ km

49. $c^2 = 11^2 + 3^2 = 121 + 9 = 130$
$c = \sqrt{130} \approx 11.402$ m

51. $c^2 = 5^2 + 5^2 = 25 + 25 = 50$
$c = \sqrt{50} \approx 7.071$ m

53. $5^2 = 4^2 + x^2 \Rightarrow x^2 = 5^2 - 4^2 = 25 - 16 = 9$
$x^2 = 9 \Rightarrow x = 3,\ A = 4^2 + \frac{1}{2} \cdot (3)(4)$
$A = 16 + 6 = 22$ in.^{2}

55. $13^2 = 12^2 + x^2 \Rightarrow x^2 = 169 - 144 = 25$
$x = 5,\ A = 4(12) + \frac{1}{2} \cdot (5)(12)$
$A = 48 + 30 = 78$ ft^2

57. $c^2 = 8^2 + 15^2 = 64 + 225 = 289$
$c = \sqrt{289} = 17$ ft

59. $c^2 = 3^2 + 4^4 = 9 + 16 = 25$
$c = \sqrt{25} = 5$ mi

61. $20^2 = 18^2 + x^2 \Rightarrow x^2 = 400 - 324 = 76$
$x = \sqrt{76} \approx 8.7$ ft

63. $A = 2(15)(20) + 2(15)(30) + 2 \cdot \frac{1}{2} \cdot (20)(12)$
$A = 1740$ ft^2

65. $C = 2 \cdot \frac{1}{2} \cdot (18)(13) \cdot 90 = \$21,060$

Cumulative Review Problems

67. $P = a + b + c = 3 + 4 + 5 = 12$ ft

69. $A = LW = 8(4) = 32$ cm^2

10.5 Exercises

1. The distance around a circle is called the circumference.

3. The diameter is two time the radius of the circle.

5. $d = 2r = 45 \Rightarrow r = 22.5$ yd

7. $d = 2r = 3.8 \Rightarrow r = 1.9$ cm

9. $C = \pi d \approx 3.14(32) \approx 100.5$ cm

11. $C = 2\pi r \approx 2(3.14)(11) \approx 69.1$ in.

13. distance = $C = \pi d \approx 3.14(28) \approx 87.9$ in.

15. distance = $5C = 5\pi d \approx 5 \cdot 3.14(24)$
distance $\approx$ 376.8 in.$\cdot \frac{1 \text{ ft}}{12 \text{ in.}} = 31.4$ ft

17. $A = \pi r^2 \approx 3.14(7)^2 \approx 153.9$ yd^2

19. $A = \pi r^2 \approx 3.14\left(\frac{44}{2}\right)^2 \approx 1519.8$ cm^2

21. $A = \pi r^2 \approx 3.14(12)^2 \approx 452.2 \text{ ft}^2$

23. $A = \pi r^2 \approx 3.14\left(\frac{90}{2}\right)^2 = 6358.5 \text{ mi}^2$

25. $A = LW + \pi r^2 \approx 120(40) + 3.14\left(\frac{40}{2}\right)^2$

$A \approx 6056 \text{ yd}^2,\ C = 6056(0.20) = \$1211.20$

27. $C = \pi d \approx 3.14(2) = 6.28 \text{ ft}$

29. distance $= 35\pi d \approx 35(3.14)(14 \cdot 2)$

distance ≈ 3077.2 in.

distance $\approx 3077.2 \text{ in.} \cdot \frac{1 \text{ ft}}{12 \text{ in.}} \approx 256.43 \text{ ft}$

31. distance $= 20{,}096 = n \cdot 2\pi r$

$20{,}096 \approx n \cdot 2(3.14)(16)$

$n \approx 200$ rev

33. $A = \pi r^2 \approx 3.14\left(\frac{64}{2}\right)^2 = 3215.36 \text{ in.}^2$

35. $A = \pi r^2 \approx 3.14\left(\frac{6}{2}\right)^2 = 28.26 \text{ ft}^2$

$\text{cost} = 28.26 \text{ ft}^2 \cdot \frac{1 \text{ yd}^2}{9 \text{ ft}^2} \cdot \frac{\$72}{\text{yd}^2} = \$226.08$

37. $C = \pi d = \pi(0.223) \approx 0.70058 \text{ m}$

39. (a) $\frac{\$6}{8 \text{ slices}} = \$0.75$ per slice

$\frac{1}{8} \cdot \pi r^2 \approx \frac{1}{8} \cdot 3.14\left(\frac{15}{2}\right)^2 \approx 22.1 \text{ in.}^2$

(b) $\frac{\$4}{6 \text{ slices}} = \$0.67$ per slice

39. (b) $\frac{1}{6} \cdot \pi r^2 \approx \frac{1}{6} \cdot 3.14\left(\frac{12}{2}\right)^2 \approx 18.8 \text{ in.}^2$

(c) 12 in.: $\frac{\$0.67}{18.8 \text{ in.}^2} \approx \$0.036 \text{ per in.}^2$

15 in.: $\frac{\$0.75}{22.1 \text{ in.}^2} \approx \$0.034 \text{ per in.}^2$

The 15 in. pizza is the better buy.

Cumulative Review Problems

41. $0.16(87) = 13.92$

43. $V = LWH = 11(5)(6) = 330 \text{ in.}^3$

Putting Your Skills to Work, Problems for Individual Investigation

1. Patio: $40(15) = 600 \text{ ft}^2$

Planter: $50(3) + 3(57) + 3(47) = 462 \text{ ft}^2$

Lawn:

$50(60) - 15(40) - 600 - 462 = 1338 \text{ ft}^2$

2. $600 \text{ ft}^2 \cdot 4 \text{ in.} \cdot \frac{1 \text{ ft}}{12 \text{ in.}} = 200 \text{ ft}^3$

3. $(50-3)(2) + (60-6) = 148 \text{ ft} \cdot \frac{12 \text{ in.}}{1 \text{ ft}}$

$= 1776 \text{ in.}$

$1776 \text{ in.} \cdot \frac{1 \text{ brick}}{8 \text{ in.}} \cdot 4 = 888 \text{ bricks}$

4. $462 \text{ ft}^2 \cdot 12 \text{ in.} \cdot \frac{1 \text{ ft}}{12 \text{ in.}} = 462 \text{ ft}^3$

5. $2(10) + 2(50) + 60 = 180$ ft of fence

Putting Your Skills to Work, Problems for Group Investigation and Cooperative Study

1. $1\text{ yd}^3 = 1\text{ yd}^3 \cdot \frac{3^3\text{ ft}^3}{\text{yd}^3} = 27\text{ ft}^3$

2. Cement:

$200\text{ ft}^3 \cdot \frac{1\text{ yd}^3}{27\text{ ft}^3} \approx 7.4\text{ yd}^3 \to 8\text{ yd}^3$

Planter:

$462\text{ ft}^3 \cdot \frac{1\text{ yd}^3}{27\text{ ft}^3} \approx 17.1\text{ yd}^3 \to 18\text{ yd}^3$

3. $A = \pi r^2 \approx 3.14\left(\frac{8}{2}+2\right)^2 = 113.04\text{ ft}^2$

$1338 - 113.04 = 1224.96\text{ ft}^2$

4. (a) Answers may vary.
(b) Answers may vary.

10.6 Exercises

1. $V = LWH$: box

3. $V = \frac{4\pi r^3}{3}$: sphere

5. $V = \frac{Bh}{3}$: pyramid

7. $V = \pi r^2 h \approx 3.14(2)^2(7) \approx 87.9\text{ m}^3$

9. $V = \frac{4\pi r^3}{3} \approx \frac{4(3.14)(4)^3}{3} \approx 267.9\text{ m}^3$

11. $V = \frac{1}{2}\cdot\frac{4\pi r^3}{3} \approx \frac{1}{2}\cdot\frac{4(3.14)(7)^3}{3}$

$V \approx 718.0\text{ m}^3$

13. $V = \frac{\pi r^2 h}{3} \approx \frac{3.14(9)^2(12)}{3} \approx 1017.4\text{ cm}^3$

15. $V = \frac{\pi r^2 h}{3} \approx \frac{3.14(5)^2(10)}{3} \approx 261.7\text{ ft}^3$

17. $V = \frac{Bh}{3} = \frac{3^2(7)}{3} = 21\text{ m}^3$

19. $V = \frac{Bh}{3} = \frac{6(12)(5)}{3} = 120\text{ m}^3$

21. $V = \pi(R^2 - r^2)h \approx 3.14(5^2 - 3^2)(20)$

$V \approx 1004.8\text{ in.}^3$

23. $\frac{\$0.09}{\text{in.}^3}\cdot\left(\frac{3.14(1)^2(6)}{3} + \frac{1}{2}\cdot\frac{4(3.14)(1)^2}{3}\right)$

$= \$0.75$

25. $V = LWH + \pi r^2 h \approx 4(3)(2) + 3.14(1)^2(2)$

$V \approx 30.28\text{ ft}^3$

27. $\text{cost} = \frac{\$4}{\text{cm}^3}\cdot\frac{3.14(5)^2(9)}{3} = \$942$

29. $V = \frac{4\pi r^3}{3} \approx \frac{4(3.14)(45{,}000)^3}{3}$

$V \approx 381{,}510{,}000{,}000{,}000\text{ mi}^3$

31. $V = \frac{4\pi r^3}{3} \approx \frac{4(3.14)(5.21)^3}{3} \approx 592.1\text{ m}^3$

33. $V = \frac{Bh}{3} = \frac{6.22(5.01)(9.212)}{3} \approx 95.7\text{ ft}^3$

35. $S = 4\pi r^2 \approx 4(3.14)(7)^2 = 615.44\text{ in.}^2$

Cumulative Review Exercises

37. $\frac{21}{40}=\frac{x}{120}\Rightarrow 40x=21(120)=2520$
$x=\frac{2520}{40}=63$

39. $2\frac{1}{4}\cdot 3\frac{3}{4}=\frac{9}{4}\cdot\frac{15}{4}=\frac{135}{16}=8\frac{7}{16}$

10.7 Exercises

1. $\frac{3}{2}=\frac{12}{n}\Rightarrow 3n=24\Rightarrow n=\frac{24}{3}=8$ m

3. $\frac{7}{2}=\frac{n}{5}\Rightarrow 2n=7(5)=35\Rightarrow n=17.5$ cm

5. $\frac{18}{5}=\frac{8}{n}\Rightarrow 18n=40\Rightarrow n=2.2$ in.

7. $\frac{9}{5}=\frac{30}{n}\Rightarrow 9n=150\Rightarrow n\approx 16.7$ m

9. $\frac{9}{12}=\frac{n}{19}\Rightarrow 12n=171\Rightarrow n=14.25$ ft

11. $\frac{n}{24}=\frac{6}{4}\Rightarrow 4n=144\Rightarrow n=36$ ft

13. $\frac{5.5}{6.5}=\frac{n}{96}\Rightarrow 6.5n=528\Rightarrow n\approx 81$ ft

15. $\frac{20}{42}=\frac{n}{8}\Rightarrow 42n=160\Rightarrow n\approx 3.8$ km

17. $\frac{n}{14}=\frac{15}{9}\Rightarrow 9n=210\Rightarrow n\approx 23.3$ cm

19. $\frac{100}{12}=\frac{P}{4}\Rightarrow 12P=400\Rightarrow P\approx 33.3$ m

21. $\frac{26}{3^2}=\frac{A}{2^2}\Rightarrow 9A=104\Rightarrow A\approx 11.6\text{ yd}^2$

Cumulative Review Problems

23. $2\cdot 3^2+4-2\cdot 5$
$=2\cdot 9+4-10=18-6=12$

25. $(5)(9)-(21+3)\div 8$
$=45-24\div 8=45-3=42$

Chapter 10 Review

1. $27\text{ ft}\cdot\frac{1\text{ yd}}{3\text{ ft}}=9\text{ yd}$

2. $2160\text{ sec}\cdot\frac{1\text{ min}}{60\text{ sec}}=36\text{ min}$

3. $90\text{ in.}\cdot\frac{1\text{ ft}}{12\text{ in.}}=7.5\text{ ft}$

4. $15{,}840\text{ ft}\cdot\frac{1\text{ mi}}{5280\text{ ft}}=3\text{ mi}$

5. $4\text{ tons}\cdot\frac{2000\text{ lb}}{1\text{ ton}}=8000\text{ lb}$

6. $15\text{ gal}\cdot\frac{4\text{ qt}}{1\text{ gal}}=60\text{ qt}$

7. $92\text{ oz}\cdot\frac{1\text{ lb}}{16\text{ oz}}=5.75\text{ lb}$

9. 59 mL = 0.059 L

10. 56 cm = 560 mm

11. 2598 mm = 259.8 cm

12. 778 mg = 0.778 g

13. 9.2 m = 920 cm

14. 7 km = 7000 m

15. 17 kL = 17,000 L

16. 473 m = 0.473 km

17. 196 kg = 196,000 g

18. 721 kg = 721,000 g

19. $\dfrac{4 \text{ L}}{24 \text{ jars}} = \dfrac{4000 \text{ mL}}{24 \text{ jars}} \approx 166.67\ \dfrac{\text{mL}}{\text{jar}}$

20. (a) $P = 7\frac{2}{3} + 4\frac{1}{3} + 5 = 17$ ft

(b) $P = 17 \text{ ft} \cdot \dfrac{12 \text{ in.}}{1 \text{ ft}} = 204$ in.

21. $42 \text{ kg} \cdot \dfrac{2.2 \text{ lb}}{\text{kg}} = 92.4$ lb

22. $20 \text{ lb} \cdot \dfrac{0.454 \text{ kg}}{\text{lb}} = 9.08$ kg

23. $15 \text{ ft} \cdot \dfrac{0.305\text{m}}{\text{ft}} \approx 4.58$ m

24. $1.8 \text{ ft} \cdot \dfrac{12 \text{ in.}}{\text{ft}} \cdot \dfrac{2.54 \text{ cm}}{\text{in.}} \approx 54.86$ cm

25. $13 \text{ oz} \cdot \dfrac{28.35 \text{ g}}{\text{oz}} = 368.55$ g

26. $F = 1.8C + 32 = 1.8(15) + 32 = 59°\text{F}$

27. $14 \text{ cm} \cdot \dfrac{0.394 \text{ in.}}{\text{cm}} \approx 5.52$ in.

28. $C = \dfrac{5F - 160}{9} = \dfrac{5(32) - 160}{9} = 0°\text{C}$

29. $200 \text{ mi} - 90\dfrac{\text{km}}{\text{hr}} \cdot 3 \text{ hr} \cdot \dfrac{0.62 \text{ mi}}{\text{km}} = 32.6$ mi

30. $\dfrac{\$0.14}{\text{oz}} \cdot 450 \text{ g} \cdot \dfrac{0.0353 \text{ oz}}{\text{g}} = \$2.22$

31. $\angle MNO$, $\angle ONM$

32. $\angle PNO$, $\angle$ONP

33. (a) $180° - 39° = 141°$

(b) $90° - 39° = 51°$

34. $m + (m + 10°) = 90° \Rightarrow 2m = 80°$

$m = \angle ABD = 40°$

35. $\angle w$, $\angle y$

36. $\angle x + \angle y = 180°$

37. $\angle z = \angle x = 65°$

$\angle w = \angle y = 180° - 65° = 115°$

38. $\angle w = \angle v = \angle u = \angle s = 45°$

$\angle t = 180° - 45° = 135°$

39. $\sqrt{64} = 8$

40. $\sqrt{\dfrac{64}{81}} = \dfrac{8}{9}$

41. $\sqrt{45} \approx 6.708$

42. $\sqrt{5} \approx 2.236$

43. $\sqrt{64}-\sqrt{49}=8-7=1$

44. $\sqrt{144}+\sqrt{9}=12+3=15$

45. $A=5.1^2+\frac{1}{2}\cdot(9.6)(5.1)\approx 50.5\text{ m}^2$

46. $13^2=12^2+a^2\Rightarrow a^2=169-144=25$
$a=5\text{ yd}$

47. $7^2=6^2+a^2\Rightarrow a^2=49-36=13$
$a=\sqrt{13}\approx 3.61\text{ ft}$

48. $C=\pi d\approx 3.14(12)\approx 37.7\text{ in.}$

49. $C=2\pi r\approx 2(3.14)(7)\approx 44.0\text{ in.}$

50. $A=\pi r^2\approx 3.14(6)^2=113.04\text{ m}^2$

51. $A=\pi r^2\approx 3.14\left(\frac{16}{2}\right)^2=200.96\text{ ft}^2$

52. $A=LW-\pi r^2\approx 12(10)-3.14(2)^2$
$A\approx 107.4\text{ ft}^2$

53. $V=\frac{4\pi r^3}{3}\approx\frac{4(3.14)(1.2)^3}{3}\approx 7.2\text{ ft}^3$

54. (a) $V=LWH-\pi r^2h$
$V\approx 4(3)(7)-3.14(1)^2(7)\approx 62.02\text{ m}^3$
(b) $\text{cost}=62.02\text{ m}^3\cdot\frac{\$1.20}{\text{m}^3}=\$72.42$

55. $V=\pi r^2h+\frac{1}{2}\cdot\frac{4\pi r^3}{3}$
$V\approx 3.14(2)^2(9)+\frac{1}{2}\cdot\frac{4(3.14)(2)^3}{3}$
$V\approx 129.787\text{ cm}^3$

56. $V=\frac{Bh}{3}=\frac{16(18)(18)}{3}=1728\text{ m}^3$

57. $V=\frac{\pi r^2h}{3}\approx\frac{3.14(17)^2(30)}{3}=9074.6\text{ yd}^3$

58. $\frac{3}{2}=\frac{45}{n}\Rightarrow 3n=90\Rightarrow n=30\text{ m}$

59. $\frac{18+5+26+5}{18}=\frac{P}{108}\Rightarrow 18P=5832$
$P=324\text{ cm}$

Chapter 10 Test

1. $145\text{ oz}=144\text{ oz}\cdot\frac{1\text{ lb}}{16\text{ oz}}+1\text{ oz}=9\text{ lb }1\text{ oz}$

2. $162\text{ g}=0.162\text{ kg}$

3. (a) $2\text{ kg}=2000\text{ g}$
(b) $2\text{ kg}\cdot\frac{2.2\text{ lb}}{\text{kg}}=4.4\text{ lb}$

4. $F=1.8C+32=1.8(200)+32=392°\text{F}$

5. $14\text{ km}+4\text{ mi}\cdot\frac{1.61\text{ km}}{\text{mi}}=20.44\text{ km}$

6. (a) $180°-41°=139°$
(b) $90°-41°=49°$

7. $m+(m+8°)=90°\Rightarrow 2m=82°$
$m=41°$

8. (a) $\angle a=\angle c=70°$
(b) $\angle d=180°-70°=110°$

9. $\angle w=\angle u=\angle t=43°$
$\angle s=\angle v=180°-43°=137°$

10. $81 = 9^2$, yes

11. $4^2 = 16,\ 5^2 = 25$
24 is not a perfect square

12. $\frac{1}{2^2} = \frac{1}{4},\ \frac{1}{3^2} = \frac{1}{9}$
$\frac{1}{8}$ is not a perfect square

13. $A = s^2 = 64 \Rightarrow s = 8$ ft

14. $\sqrt{49} = 7$

15. $\sqrt{\frac{9}{16}} = \frac{3}{4}$

16. $\sqrt{6} \approx 2.45$

17. $\sqrt{25} + \sqrt{36} = 5 + 6 = 11$

18. $A = \frac{1}{2}bh = \frac{1}{2}(10)(12) = 60 \text{ cm}^2$

19. $A = 30(20) + \frac{1}{2}(20)(10) = 700 \text{ ft}^2$

20. $7^2 = 3^2 + a^2 \Rightarrow a^2 = 40 \Rightarrow a = \sqrt{40}$
$a \approx 6.32$ cm

21. $c^2 = 14^2 + 11^2 = 317 \Rightarrow c = \sqrt{317}$
$c \approx 17.8$ ft

22. $C = \pi d \approx 3.14(5.15) = 16.171$ m

23. $2\pi R = n\pi r \Rightarrow 2(2.1) = n(0.75)$
$n = 5.6$ rev

24. $A = \pi r^2 \approx 3.14(1.2)^2 \approx 4.5 \text{ cm}^2$

25. $\frac{\$20}{\text{yd}^2} \cdot 3.14(3 \text{ ft})^2 \cdot \frac{\text{yd}^2}{9 \text{ ft}^2} = \$62.80$

26. $V = \pi r^2 h \approx 3.14(3.4)^2(7.1) \approx 257.72 \text{ cm}^3$

27. $V = \frac{4\pi r^3}{3} \approx \frac{4(3.14)(8)^3}{3} \approx 2143.57 \text{ in.}^3$

28. $V = \frac{\pi r^2 h}{3} \approx \frac{3.14(8)^2(12)}{3} = 803.84 \text{ cm}^3$

29. $V = \frac{Bh}{3} = \frac{10(7)(12)}{3} = 280 \text{ m}^3$

30. $\frac{4}{13} = \frac{18}{n} \Rightarrow 4n = 234 \Rightarrow n = 58.5$ in.

31. $\frac{h}{2.5} = \frac{20}{2} \Rightarrow 2h = 50 \Rightarrow h = 25$ ft

32. $\frac{1}{4.5} = \frac{W}{36} \Rightarrow 4.5W = 36 \Rightarrow W = 8$ cm

Cumulative Test for Chapters 1-10

1. $0 > -5$

2. $0.5 > 0.15$

3. $-8 > -18$

4. $8 < 116$

5. $|-7| = 7,\ |-9| = 9 \Rightarrow |-9| > |-7|$

6. $|-6| = 6,\ |2| = 2 \Rightarrow |-6| > |2|$

7. (a) $2(9) + 2(12) = 42$ ft
(b) $2.5(42) = \$105$

7. (c) $9(12) = 108 \text{ ft}^2$

8. $12 - 5^2 + 3 = 12 - 25 + 3 = -13 + 3 = -10$

9. $4(8 - 12) = 4(-4) = -16$

10. $-15(-3) = 45$

11. $\frac{1}{3}\left(\frac{2}{5}\right) \cdot \frac{2}{2} + \frac{1}{2} \cdot \frac{15}{15} = \frac{4+15}{30} = \frac{19}{30}$

12. $\frac{-3}{4} \div \frac{7}{8} = -\frac{3}{4} \cdot \frac{8}{7} = -\frac{6}{7}$

13. $2x + 3x^2 - 8 + x - 17 = 3x^2 + 3x - 25$

14. $x - 20 = 18 \Rightarrow x = 18 + 20 = 38$

check: $38 - 20 \overset{?}{=} 18,\ 18 = 18$

15. $x + 3 = -12 \Rightarrow x = -12 - 3 = -15$

check: $-15 + 3 \overset{?}{=} -12,\ -12 = -12$

16. $3 - 8 + x = 14 - 9 \Rightarrow -5 + x = 5 \Rightarrow x = 10$

check: $3 - 8 + 10 \overset{?}{=} 14 - 9,\ 5 = 5$

17. $\frac{x}{6} = 4 \Rightarrow x = 6(4) = 24$

check: $\frac{24}{6} \overset{?}{=} 4,\ 4 = 4$

18. $5x - 9(2) = 48 \Rightarrow 5x = 48 + 18 = 66$

$x = \frac{66}{5} = 13.2$

check: $5(13.2) - 9(2) \overset{?}{=} 48,\ 48 = 48$

19. $4(x + 3) = 4x + 4(3) = 4x + 12$

20. $(x + 2)(x - 1) = x^2 - x + 2x - 2$
$= x^2 + x - 2$

21. $(4x + 3y - 9) + (-3x + 7y + 6)$
$= 4x + 3y - 9 - 3x + 7y + 6$
$= x + 10y - 3$

22. $0.83(250) = \$207.50$

23. $\frac{26{,}275 - 23{,}450}{23{,}450} \approx 0.120469 \rightarrow 12.05\%$

24. $\frac{76 + 85 + 73 + 75 + 81}{5} = 78$

25. 73, 75, 76, 81, 85

median = 76

26. (a) $y = 2x + 1$
$y = 2(-1) + 1 = -1 \Rightarrow (-1, -1)$
$y = 2(0) + 1 = 1 \Rightarrow (0, 1)$
$y = 2(1) + 1 = 3 \Rightarrow (1, 3)$

(b)

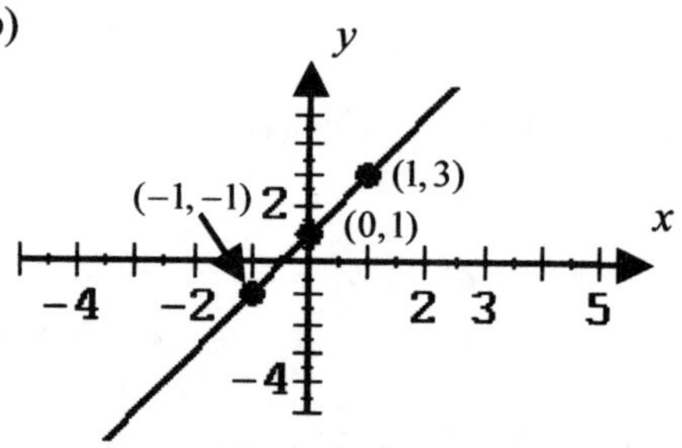

27. (a) $96 \text{ in.} \cdot \frac{1 \text{ ft}}{12 \text{ in.}} = 8 \text{ ft}$

(b) $2.25 \text{ lb} \cdot \frac{16 \text{ oz}}{\text{lb}} = 36 \text{ oz}$

28. (a) 57 cm = 570 mm
(b) 4216 mL = 0.004216 kL

29. (a) $280 \text{ gal} \cdot \frac{3.79 \text{ L}}{\text{gal}} = 1061.2 \text{ L}$

(b) $18 \text{ m} \cdot \frac{1.09 \text{ yd}}{\text{m}} = 19.62 \text{ yd}$

(c) $186 \text{ g} \cdot \frac{0.0353 \text{ oz}}{\text{g}} \approx 6.57 \text{ oz}$

30. $\angle ZXY$, $\angle YXZ$

31. $\angle MNO$ is called a straight angle.

32. $\angle MNP$ is called an acute angle.

33. $\angle ONP$ is called an obtuse angle.

34. $\angle e = \angle g = \angle i = 35°$
$\angle f = \angle h = 180° - 35° = 145°$

35. $\sqrt{144} = 12$

36. $\sqrt{\frac{81}{100}} = \frac{9}{10}$

37. $\sqrt{10} \approx 3.2$

38. $c^2 = 6^2 + 8^2 = 36 + 64 = 100 \Rightarrow c = 10$

39. $C = \pi d \approx 3.14(14) = 43.96$ in.

40. $V = \pi r^2 h \approx 3.14(2.7)^2(6.3) \approx 144.21 \text{ cm}^3$

41. $V = \frac{Bh}{3} = \frac{14(10)(12)}{3} = 560 \text{ in.}^3$

42. $\frac{h}{20} = \frac{15}{10} \Rightarrow 10h = 300 \Rightarrow h = 30$ ft

Practice Final Examination

1. 7,543,876
= 7,500,000 nearest hundred thousand

2. A number decreased by 5: $x - 5$

3.
$$\begin{aligned}(2+5+x)+4 &= (7+x)+4 \\ &= (x+7)+4 \\ &= x+(7+4) \\ &= x+11\end{aligned}$$

4. $x + y\big|_{x=9,y=12} = 9 + 12 = 21$

5. $6+5+15+4+(15+6)+(5+4) = 60$ ft

6.
$$\begin{array}{r} 751 \\ \underline{-482} \\ 269 \end{array}$$

check: $269 + 482 \stackrel{?}{=} 751$, $751 = 751$

7.
$$\begin{array}{r} 5282 \\ \underline{\times 806} \\ 31692 \\ \underline{42256} \\ 4257292 \end{array} \rightarrow 4,257,292$$

8. $812,869 \div 743 = 1094$ R27
$$\begin{array}{r} 1094 \\ 743\overline{)812869} \\ \underline{743} \\ 6986 \\ \underline{6687} \\ 2999 \\ \underline{2972} \\ 27 \end{array}$$

9. $7+15\div 3-3^2=7+5-9=12-9=3$

10. $y+4y+3y=16 \Rightarrow 8y=16 \Rightarrow y=2$

11. $30+30+70+190=\$320$

12. $-5<-3$

13. $-|-13|=-(-(-13))=-(13)=-13$

14. $7+(-3)+9+(-4)=4+5=9$

15. $-9-7+8=-16+8=-8$

16. $(-6)(2)(-1)(5)(-3)=(-12)(-5)(-3)$
$=60(-3)=-180$

17. $64\div(-8)=-8$

18. $-9+5y+7-3y=5y-3y-9+7$
$=2y-2$

19. $a^2-b\big|_{a=-4,b=7}=(-4)^2-7=16-7=9$

20. $-3(a-2)=-3a-(-3)(2)=-3a+6$

21. $5(3-7)=x-3 \Rightarrow 5(-4)=x-3$
$x-3=-20 \Rightarrow x=-20+3=-17$

check: $5(3-7)\stackrel{?}{=}-17-3,\ -20=-20$

22. $6(3x)=54 \Rightarrow 18x=54 \Rightarrow x=3$

check: $6(3(3))\stackrel{?}{=}54,\ 54=54$

23. (a) $L=5W$

(b) $40=5W \Rightarrow W=\dfrac{40}{5}=8$ ft

24. $21(10)+(21-14)(9)+7(11)=350 \text{ in.}^2$

25. $A=bh=b(9)=108 \Rightarrow b=12$ m

26. $V=10W(5)=200 \Rightarrow W=\dfrac{200}{50}=4$ ft

27. (a) $(21 \text{ ft})(15 \text{ ft})\cdot\dfrac{1 \text{ yd}^2}{9 \text{ ft}^2}=35 \text{ yd}^2$

(b) $35 \text{ yd}^2\cdot\dfrac{\$18}{\text{yd}^2}=\$630$

28. $(5y)(z^6)(5y^4)(3z^2)=5\cdot 5\cdot 3y^{4+1}z^{6+2}$
$=75y^5z^8$

29. $V=LWH=(2y^4)(3y^2)(5y)=30y^7$

30. $\dfrac{57}{7}=8\dfrac{1}{7}$

31. $3\dfrac{1}{5}=\dfrac{5\cdot 3+1}{5}=\dfrac{15+1}{5}=\dfrac{16}{5}$

32. $\dfrac{72\cancel{y}}{117\cancel{y}z}=\dfrac{\cancel{9}\cdot 8}{\cancel{9}\cdot 13z}=\dfrac{8}{13z}$

33. $\dfrac{27}{-45}=-\dfrac{\cancel{9}\cdot 3}{\cancel{9}\cdot 5}=-\dfrac{3}{5}$

34. $\dfrac{18x^7}{42x^{10}}=\dfrac{\cancel{6}\cdot 3\cancel{x^7}}{\cancel{6}\cdot 7\cancel{x^7}\cdot x^3}=\dfrac{3}{7x^3}$

35. $(3^3)^2=3^{3\cdot 2}=3^6=729$

36. $\dfrac{21}{49}=\dfrac{\cancel{7}\cdot 3}{\cancel{7}\cdot 7}=\dfrac{3}{7}$

37. $\dfrac{156 \text{ ft}}{12 \text{ hr}}=13$ ft per hr

38. (a) $\frac{42}{14} = 3$ secretaries per lawyer

(b) $\frac{12}{3} = 4$ paralegals per lawyer

(c) $\frac{12}{3} = \frac{n}{80} \Rightarrow 3n = 960$

$n = 320$ paralegals

39. $\frac{24}{x} = \frac{8}{5} \Rightarrow 8x = 24(5) = 120 \Rightarrow x = 15$

check: $\frac{24}{15} \stackrel{?}{=} \frac{8}{5}, \ \frac{8}{5} = \frac{8}{5}$

40. $\frac{1}{90} = \frac{5}{n} \Rightarrow n = 5(90) = 450$ mi

41. $\frac{\cancel{8}y}{\cancel{15}} \cdot \frac{\overset{2}{\cancel{30}}}{\underset{3}{\cancel{24}}y^2} = \frac{2}{3y}$

42. $\frac{9}{14} \div \frac{-72}{21} = -\frac{\cancel{9}}{\cancel{7} \cdot 2} \cdot \frac{\cancel{7} \cdot 3}{\cancel{9} \cdot 8} = -\frac{3}{16}$

43. $\frac{1}{3} \div 4 = \frac{1}{3} \cdot \frac{1}{4} = \frac{1}{12}$ lb in each container

44. $6x = 2 \cdot 3 \cdot x, \ 18x = 2 \cdot 3^2 x, \ 27 = 3^3$

$LCM = 2 \cdot 3^3 x = 54x$

45. $\frac{8}{y} - \frac{4}{y} = \frac{8-4}{y} = \frac{4}{y}$

46. $\frac{-5}{16} \cdot \frac{3}{3} + \frac{9}{24} \cdot \frac{2}{2} = \frac{-15+18}{48} = \frac{3}{48} = \frac{1}{16}$

47. $\frac{5x}{18} \cdot \frac{5}{5} - \frac{7x}{45} \cdot \frac{2}{2} = \frac{25x - 14x}{90} = \frac{11x}{90}$

48. $12\frac{1}{7} \cdot \frac{3}{3} - 7\frac{13}{21} = 11\frac{24}{21} - 7\frac{13}{21} = 4\frac{11}{21}$

49. $3\frac{3}{7} \div 6\frac{2}{3} = \frac{24}{7} \cdot \frac{3}{20} = \frac{6 \cdot \cancel{4} \cdot 3}{7 \cdot \cancel{4} \cdot 5} = \frac{18}{35}$

50. $\left(\frac{3}{4}\right)^2 + \frac{1}{9} \div \frac{1}{3} = \frac{9}{16} + \frac{1}{9} \cdot \frac{3}{1}$

$= \frac{9}{16} \cdot \frac{3}{3} + \frac{1}{3} \cdot \frac{16}{16}$

$= \frac{27+16}{48} = \frac{43}{48}$

51. $\frac{3}{4}x = 28 \Rightarrow x = \frac{4 \cdot 28}{3} = \frac{112}{3}$

52. $1\frac{1}{3} - \frac{6}{9} = \frac{4}{3} \cdot \frac{3}{3} - \frac{6}{9} = \frac{12-6}{9} = \frac{6}{9} = \frac{2}{3}$ in.

53. $(-4y^2 - 5) - (2y^2 + 3y + 2) + (-y^2 - 3)$

$= -4y^2 - 5 - 2y^2 - 3y - 2 - y^2 - 3$

$= -7y^2 - 3y - 10$

54. $(-3x)(x-7) = -3x^2 + 21x$

55. $(z^5)(z^6 - 3z - 5) = z^{11} - 3z^6 - 5z^5$

56. $(y+7)(4y^2 - 4y + 6)$

$= 4y^3 - 4y^2 + 6y + 28y^2 - 28y + 42$

$= 4y^3 + 24y^2 - 22y + 42$

57. $\angle b = x, \ \angle a = x + 40°, \ \angle c = 3x$

58. $(y+5)(y-3) = y^2 - 3y + 5y - 15$

$= y^2 + 2y - 15$

59. $(x-4)(4x+3)=4x^2+3x-16x-12$
$=4x^2-13x-12$

60. $4z+12x=4(z+3x)$

61. $9y^5z^7+3y^6z^3=3y^5z^3(3z^4+y)$

62. $2x^2y^2-8xy+12x^2y=2xy(xy-4+6x)$

63. $-15-24=2x+15-3x \Rightarrow -39=-x+15$
$x=15+39=54$

check: $-15-24\overset{?}{=}2(54)+15-3(54)$
$-39=-39$

64. $-\frac{16}{4}=-3(6z)-5z \Rightarrow -4=-18z-5z$

$23z=4 \Rightarrow z=\frac{4}{23}$

check: $-\frac{16}{4}\overset{?}{=}-3\left(6\cdot\frac{4}{23}\right)-5\cdot\frac{4}{23}$
$-4=-4$

65. $-z=\frac{1}{4} \Rightarrow z=-\frac{1}{4}$

check: $-\left(-\frac{1}{4}\right)\overset{?}{=}\frac{1}{4},\ \frac{1}{4}=\frac{1}{4}$

66. $5y-9-9y=23 \Rightarrow -4y=32 \Rightarrow y=-8$

check: $5(-8)-9-9(-8)\overset{?}{=}23,\ 23=23$

67. $7a-5+3a=-9+6a$
$10a-5=-9+6a \Rightarrow 4a=-4 \Rightarrow a=-1$

check: $7(-1)-5+3(-1)\overset{?}{=}-9+6(-1)$
$-15=-15$

68. $-2(y-5)=6(y+5)-12$
$-2y+10=6y+30-12=6y+18$
$-8y=8 \Rightarrow y=-1$

69. $\frac{x}{2}+\frac{x}{3}=10 \Rightarrow 3x+2x=60$
$5x=60 \Rightarrow x=12$

70. (a) $L=$ fall students
$L+95=$ spring students
$L-75=$ summer students

(b) $L+(L+95)+(L-75)=386$

(c) $3L+20=386 \Rightarrow 3L=366$
$L=122$ fall students
$L+95=217$ spring students
$L-75=47$ summer students

(d) $122+(122+95)+(122-75)\overset{?}{=}386$
$386=386$

71. $751.7596=751.76$ nearest hundredth

72. $0.588+75+8.59=84.178$

73. $0.042\times 0.04=0.00168$

74. $40(9.50)+12(13.65)=\$543.80$

75. $-4.75\div 3.2=-1.484375$
$=-1.48$ nearest hundredth

76.

Fraction	Decimal	Percent
$\frac{1}{4}$	0.25	25%
$\frac{14}{25}$	0.56	56%
$3\frac{1}{2}$	3.5	350%

77. $79 = 0.25n \Rightarrow n = \dfrac{79}{0.25} = 316$

78. $0.30(90) = 27$

79. $\dfrac{5.75}{28.50} \approx 0.20175$
$= 20\%$ nearest hundredth

80. $100\% - 60\% = 40\%$

81. (a) $7(50) - 6(50) = 50$ students
(b) $4(50) + 7(50) = 550$ students

82. (a) $40\% + 30\% = 70\%$
(b) $0.10(2600) = \$260$

83.

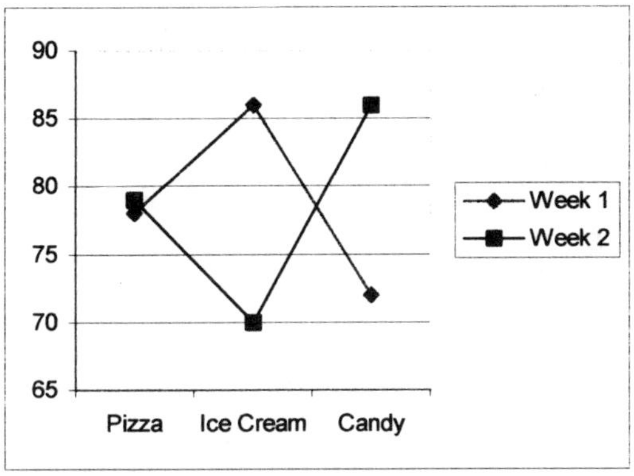

84. 59, 69, 72, 77, 78, 83, 87, 92, 95
median = 78

85. $\dfrac{89+87+92+97+85+85+95}{7} = 90°\text{F}$

86. 2, 7, 7, 8, 8, $\underbrace{9,\ 9,\ 9,}_{\text{mode}}$ 10

87. (a) $R(-3,5)$
(b) $T(0,3)$
(c) $S(-4,-4)$
(d) $U(6,2)$

88. $y = 2x - 4 \Rightarrow x = \dfrac{y+4}{2}$

$x = \dfrac{-6+4}{2} = -1 \Rightarrow (-1,-6)$

$x = \dfrac{-4+4}{2} = 0 \Rightarrow (0,-4)$

$x = \dfrac{-8+4}{2} = -2 \Rightarrow (-2,-8)$

89. $y = 3x - 2$

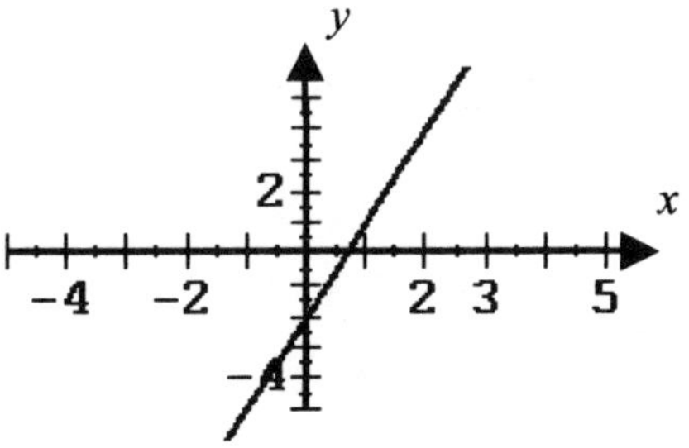

90. $y = -3$

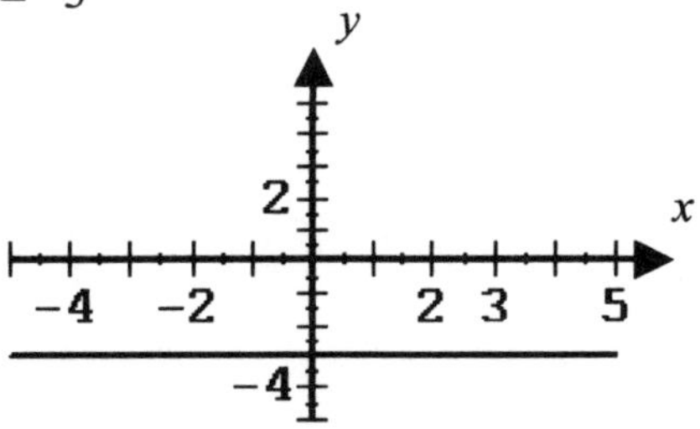

91. $14 \text{ gal} \cdot \dfrac{4 \text{ qt}}{\text{gal}} = 56 \text{ qt}$

92. $8 \text{ km} = 8000 \text{ m}$

93. $13 \text{ ft} \cdot \dfrac{0.305 \text{ m}}{\text{ft}} = 3.97 \text{ m}$

94. $\sqrt{81} = 9$

95. $\sqrt{121} + \sqrt{16} = 11 + 4 = 15$

96. $26^2 = 24^2 + x^2 \Rightarrow x^2 = 676 - 576 = 100$
$x = 10$

97. $C = \pi d \approx 3.14(13) = 40.82$ ft

98. $A = \pi r^2 \approx 3.14\left(\frac{18}{2}\right)^2 = 254.34 \text{ m}^2$

99. $V = \frac{Bh}{3} = \frac{8(10)(9)}{3} = 240 \text{ ft}^3$

100. $\frac{7}{3} = \frac{49}{n} \Rightarrow 7n = 3(49) \Rightarrow n = 21$ m